PROGRESS IN COLLOID & POLYMER SCIENCE

Editors: H.-G. Kilian (Ulm) and G. Lagaly (Kiel)

Volume 86 (1991)

Progress in Analytical Ultracentrifugation

Guest Editor: W. Borchard (Duisburg)

Springer-Verlag Berlin Heidelberg GmbH

ISBN 978-3-662-15686-5   ISBN 978-3-7985-1683-0 (eBook)
DOI 10.1007/978-3-7985-1683-0
ISSN 0340-255 X

© Springer-Verlag Berlin Heidelberg 1991
Originally published by Dr. Dietrich Steinkopff Verlag GmbH & Co. KG, Darmstadt in 1991
Softcover reprint of the hardcover 1st edition 1991
Chemistry editor: Dr. Maria Magdalene Nabbe; English editor: James Willis; Production: Holger Frey.

Type-Setting: Graphische Texterfassung, Hans Vilhard, D-6126 Brombachtal

Preface

Since the first symposium, which was organized by D. Riesner at the Technische Hochschule Darmstadt in 1978 six further meetings have taken place in Germany. Up to now the aim was to join different teams working in various fields of analytical ultracentrifugation. The field includes general theory, applications for basic research problems in biochemistry, biophysical chemistry, macromolecular chemistry and colloid chemistry.

As a modern analytical instrument has to be equipped with on-line registration techniques concerning all measurable quantities like for instance optical properties in the visible, IR- and UV-range this development has been advanced mainly by different groups in the last years. This includes also the possibilities to work with multichannel rotors to broaden the use of this important analytical method especially in the industrial application. Although there is a need for analytics of macromolecules since long time an universal modern analytical ultracentrifuge is not yet on the market. The conference at Duisburg was characterized by the opening of a mainly European conference to all users in the world thus enabling informative and stimulating discussions and the enhancement of contacts between different groups of various countries. Thanks to the efforts of the organizers the number of about 20 participants in the least years could be increased up to 60 despite the travel restrictions due to the Gulf War.

The contributions from the recent conference are opened by a historical paper of one of the pioneers in ultracentrifugation and close up with an outlook for future analytical requirements. At this meeting the emphasis of applications of ultracentrifugational techniques is in the field of biochemistry. Therefore we are glad that Prof. Dr. D. Schubert agreed to write a survey article concerning this area which is the first contribution of the biochemical section. A second section is devoted to technical developments of detection systems and progress in the analytical application to synthetic polymers. The last section concerns ultracentrifugation of gels and emulsions.

The 7th symposium was kindly sponsored by Bayer AG (Leverkusen), Henkel KG aA (Düsseldorf), Hüls AG (Marl) and especially by Beckman Instruments GmbH (München) by which the newest analytical ultracentrifuge was presented at Duisburg. The forthcoming of this issue was made possible by the enduring assistance of Dipl.-Chem. Helmut Cölfen.

W. Borchard (Duisburg)

Contents

Progress in Colloid & Polymer Science   Progr Colloid Polym Sci 86:1—11 (1991)

The contribution of the analytical ultracentrifuge to the technology and development of lipoprotein research, 1948—1991

J. R. Orr*, E. F. Dowling, and F. T. Lindgren

Lawerence Berkeley Laboratory, University of California, Donner Laboratory, Berkeley, California, USA

Abstract: Application of the analytic ultracentrifuge to lipoproteins began in 1935 with McFarlane's studies of normal and pathological serum, by which were unexplained multicomponents and distortions in the region of the albumin boundary. In 1948, Gofman and coworkers started their studies at Donner Laboratory with the newly developed electrically driven Spinco analytic ultracentrifuge designed by Pickels. In their work, there usually developed, after prolonged ultracentrifugation of undiluted sera, an actual dip below the baseline. Raising the serum density to 1.063 g/ml resulted in the elimination of the dip phenomenon and the density sensitive "X protein" floated, giving an inverse peak as predicted. The past 43 years have seen great development in all aspects of lipoprotein study. Much of this was made possible by the analytic and preparative ultracentrifuges developed by Pickels. — At Donner Laboratory, ultracentrifugal flotation techniques have been developed and refined over the past 35 years. Important technological details include the fabrication of improved analytical cells and special offset centerpieces. The manufacturing of epoxy molds and the fabrication of double-sectored centerpieces are illustrated and described. To eliminate rotations of the cell during acceleration, cell housings are slotted and rotors are pinned for precise radial alignment. All phases of data acquisition and E machine control are accomplished by a micro-computer. Tracings of the Schlieren plots are processed by a VAX station 3200 with almost instantaneous numerical and graphical output.

Key words: Analytical ultracentrifugation; lipoproteins; history; technology; analytic cell fabrication

The early history of lipoprotein isolation and characterization dates back more than 60 years, and it is beneficial to describe some of the development that occurred before about 1965, especially for the benefit of younger scientists, since many libraries no longer have access to these earlier journals, publications, and reprints.

Although the first isolation of a lipoprotein was done by Macheboef [1] in 1928, there were no physical-chemical characterizations of this lipoprotein at that time. It was called "ceonapse precipitated by acid" or CA, obtained by precipitation from horse serum using half-saturated ammonium sulfate. After the development of electrophoresis by Tiselius [2], Macheboef reported this CA to be an *a*

globulin, and we now know this to be the first isolated high-density lipoprotein (HDL).

A most important development for characterization of lipoproteins was the invention, in 1924, of the ultracentrifuge by Svedberg and Rinde [3]. In 1927, Svedberg and Lysholm [4] developed a higher speed ultracentrifuge with an optical system to observe the migration of protein boundaries. Further developments of the oil turbine high-speed drive and the Lamm scale optical system [5] led to serum protein studies. In 1935, McFarlane [6] studied normal and pathological human serum; some of the latter contained elevated cholesterol and other lipids. Figure 1 shows examples of these first analytic ultracentrifuge (AnUC) plots. Of in-

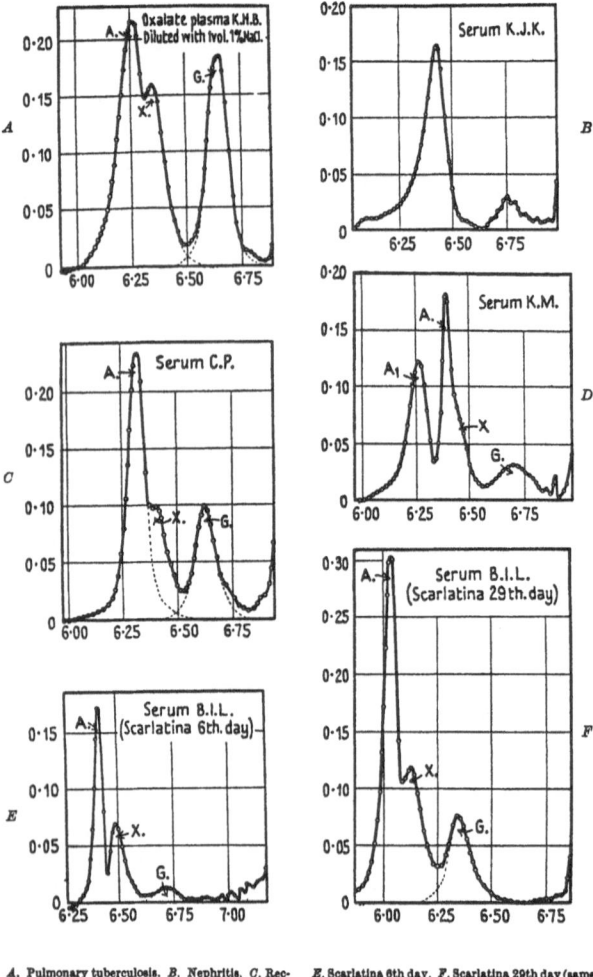

A. Pulmonary tuberculosis. B. Nephritis. C. Rectal carcinoma. D. Malignant tumor of bile duct. E. Scarlatina 6th day. F. Scarlatina 29th day (same person).

(McFarlane, 1935, c.)

XBL 8910-3766

Fig. 1. Early sedimentation patterns by McFarlane showing the apparent labile nature of the "X protein". (Reprinted with permission from Ann NY Acad Sci (1980) 348:1—15.)

terest was the presence of a density- and time-sensitive component in the region of the albumin boundary. Because of its labile nature, influenced by time, salt concentration and plasma dilution, it was called the "X protein". Others, such as von Mutzenbecker and Pederson [7] later verified these puzzling observations, but were unable to explain these anomalies. During World War II a group at Harvard worked on blood plasma fractionation and plasma substitutes, and Cohen, Oncley, Edsall, and their group isolated two different lipid-containing fractions from human plasma by low-temperature, low-salt ethanol precipitation. One was a dense (1.10 g/mL) a lipoprotein, and the other a high

molecular weight (1.3 × 10⁶ Daltons), low density (1.03 g/mL) β lipoprotein [8].

Meanwhile, Pederson made extensive AnUC studies of human and animal serum, published as a book [9]. These results suggested that the β lipoprotein was the troublesome, density-sensitive "X protein" and all studies for over a decade had been unable to characterize this low-density lipoprotein (LDL) and explain its behavior in the AnUC of plasma and serum.

Resolution of the "X protein" by AnUC lipoprotein flotation

Some 43 years ago, John Gofman began to study the process of atherosclerosis at Donner Laboratory, University of California, Berkeley, at first with a few graduate students. His objective was to determine how cholesterol and other lipids were carried in the blood stream. At this time the first commercially available analytic ultracentrifuge (AnUC) was obtained from the Specialized Instrument Company, which later became part of the Beckman Corporation. This AnUC was designed by Pickels while at the Rockefeller Institute and was equipped with an electric drive and with the new continuous dn/dx Thovert Philpot-Svensson optical system [10]. This new optical system eliminated the tedious manual plots required by the Lamm scale method [5] that had been one of the technical problems earlier workers had had to contend with. These anomalies in the neighborhood of the albumin peak seemed to increase in severity with time, and many samples exhibited a dip below baseline, something not explained by traditional multicomponent analysis (see Fig. 2). Considering the β-lipoprotein reported by the Harvard group, which had a low density of approximately 1.03 g/mL (slightly less than the density of whole plasma or serum), suddenly an important question arose: does the macromolecular β-lipoprotein (or "X protein") see the density of the serum small molecular background, i.e., 1.0063 g/mL or does it see the density of the serum background *plus* the density increment of the much smaller serum proteins such as albumin? If the latter were the case, the β-lipoprotein would ultimately accumulate on the radial side of the albumin boundary with time, and we conceived of a "pile-up" analysis to explain the distortions. Figure 3 shows how this analysis could result in a time-dependent dip below the baseline. If the pile-up were roughly

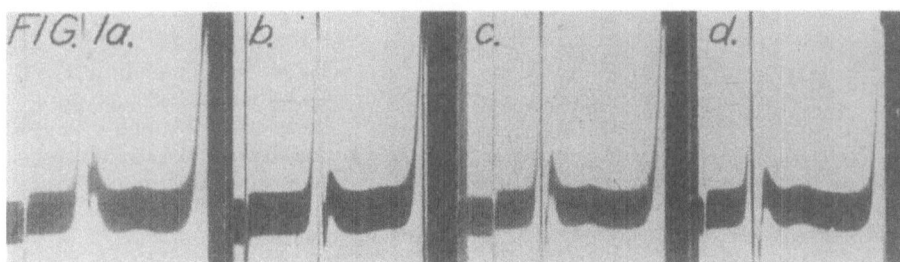

Fig. 2. Donner reproduction of albumin boundary distortions and the "dip phenomenon". (Reprinted with permission from Ann NY Acad Sci (1980) 348:1—15.)

Gaussian, then the Schlieren optical system detecting dn/dx would result in a biphasic pattern that would be superimposed on the main albumin peak. If this pile-up occurred at different regions of the albumin boundary, all the bizarre anomalies could be explained. As our AnUC run was almost finished, we shut it down, took another plasma sample aliquot and raised the salt background density with NaCl to about twice that of whole plasma, i.e., 1.063 g/ml. Eagerly, we waited for pump-down and to get the rotor and sample up to speed. Figure 4 shows the first flotation of the elusive "X protein" in the presence of whole serum. The rising inverse peak, essentially of total LDL, was confirmed and the slowly sedimenting albumin peak was symmetrical, without a trace of distortion or anomaly. Repeating this experiment many times and confirming that the area over the peak was related to lipoprotein concentration, we decided to write our first paper.

We eagerly wrote up this first paper and submitted it to the *Journal of Biological Chemistry*, expecting acceptance of this new interpretation and confirmatory data. However, after weeks passed we received the unbelievable news that the paper had been rejected by both reviewers. One reviewer thought we were confusing our interpretation with the Ogston-Johnston [11] anomaly. The other felt we were inexperienced in the field, and since several ultracentrifugal experts had come to consistent yet different conclusions, our manuscript could not be published until we had more definitive conclusions. We were frustrated and disappointed, but continued our arguments by correspondence with the editor. Finally, thanks to our perseverance and the scientific understanding of John Edsall, then the editor, our manuscript was accepted. In June of 1949, our first paper [12] was published and it was the only plasma lipoprotein paper published that year. Since 1927, there had been only four plasma lipoprotein papers published [1, 6, 7, 8]. Thus, with this landmark paper the decade-long mystery of the

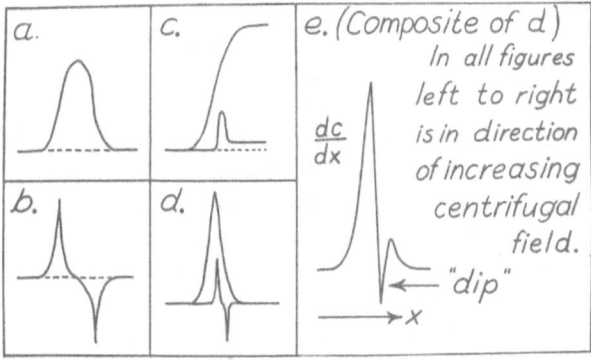

Fig. 3. Analysis of the pile-up hypothesis that could explain the dip phenomenon. (Reprinted with permission from Ann NY Acad Sci (1980) 348:1—15.)

elusive "X protein" was resolved and a new era of characterization and quantification of lipoproteins by AnUC flotation began.

While awaiting publication of our first paper, we began to apply the new technique of lipoprotein flotation to normal and cholesterol-fed rabbits, as was first done by Anitschkow [13]. Gofman et al. studied normal humans and patients with proven cardiovascular disease. The paper was sent to *Science* and quickly accepted and published, in marked contrast with our first paper. These findings indicated that, in the rabbit developing atherosclerosis, there was a minimal increase in LDL but a marked increase of the cholesterol-rich, S_f^0 10—30-class lipoproteins (see Fig. 5). According to present nomenclature, these would be described as intermediate density lipoproteins (IDL) and the smaller, higher density class of very low density lipoprotein (VLDL). These studies further showed coronary patients, when compared to normal controls, had elevated S_f^0 12—20, or IDL. The general features of these findings are shown in Fig. 5. Thus, the early studies at Donner had identified an "atherogenic class" of low density lipoproteins by

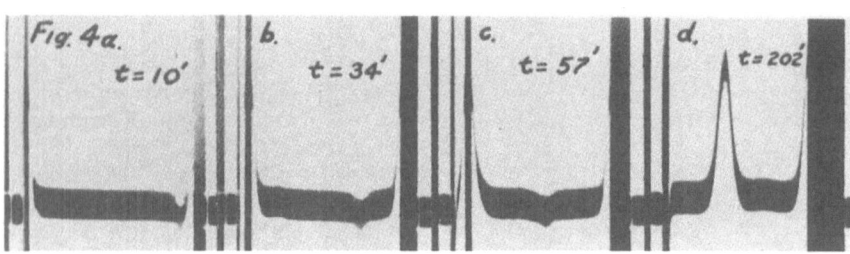

Fig. 4. First flotation of the β lipoproteins (LDL) as an inverse peak obtained by raising the density to 1.063 g/ml. (Reprinted with permission from Ann NY Acad Sci (1980) 348:1—15.)

quantitative AnUC flotation which would alert and stimulate the scientific community that was interested in coronary artery disease (CHD). After these two initial papers, other scientists, at first notably at NIH, began to isolate by ultracentrifugation, to characterize, and to study plasma lipoproteins. These included the early Donner studies by Shore, Nichols, and Freeman [15] on "clearing factor" identified as a lipolytic mechanism and analogous studies at NIH by Brown, Boyle, and Anfinson [16] that identified the enzymatic transformation of chylomicrons and VLDL by what is now called "lipoprotein chylomicron lipase". Two years later, and almost simultaneously, three groups: Havel et al. [17] (NIH), Hillyard et al. [18] (UC Berkeley), and Lindgren et al. [19] (Donner Laboratory) published, with slight differences, sequential flotation and isolation of all the major plasma lipoprotein classes. The latter study [19] also identified by AnUC the basic nature of the lipoprotein transformations induced by IV-heparin injection.

At first, the lipid composition of each major lipoprotein class was studied and found to contain all the major lipids in different proportions — triglycerides, phospholipids, free cholesterol, cholesteryl esters, and small amounts of free fatty acids. The other variable component was peptide, varying from some 2% for chylomicrons, 10—15% for VLDL, 25% for LDL, and as much as 55% for HDL. Little was known about the nature of the peptide moiety, but by about 1956, studies at both NIH by Avigan et al. [21] and at Donner by Shore [21] began to characterize the major peptides of VLDL, LDL, and HDL by their specific N-terminal amino acids. The concept evolved that the nature of the peptide determined what kind of a lipoprotein might be synthesized, say in the liver, organs or other sites. Soon, analyses became more sensitive and quantitative and it became clear that the occur-

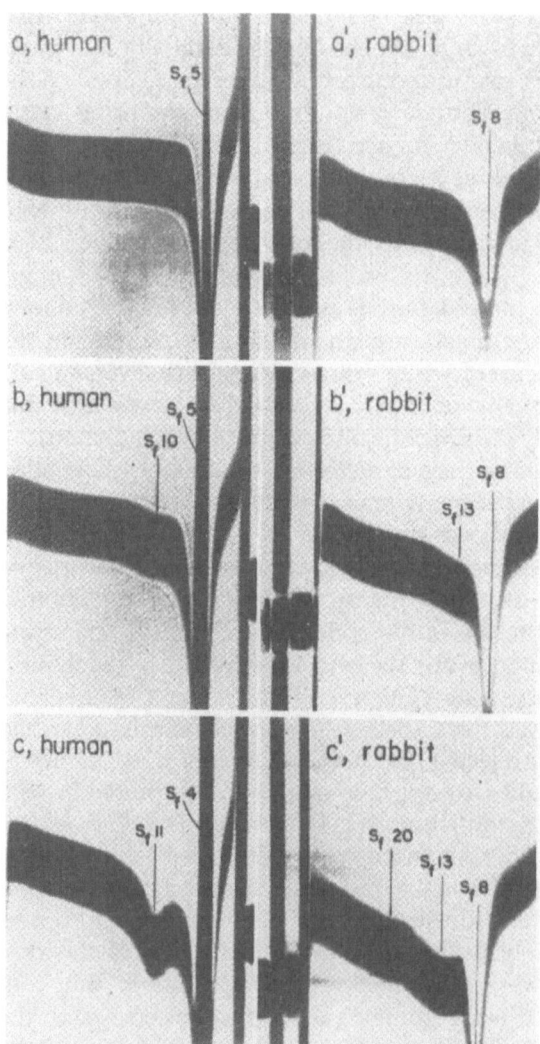

Fig. 5. Ultracentrifugal flotation diagrams of the normal rabbit (a') and the rabbit pattern developing hypercholesterolemia and atherosclerosis (b',c'). Analogous human patterns are shown to the left (a, b, and c). (Reprinted with permission from Science (copyright 1950 by the AAAS): Gofman et al. J Biol Chem (1949) 179:973.)

rence of peptides was more complicated than, say, the glutamic-N-terminal for LDL and VLDL, and aspartic-N-terminal for HDL. Both N-terminal and C-terminal amino acids were characterized in sub-fractions of the major lipoprotein classes. The first solubilization of apo-HDL was achieved by Scanu et al. [22], and later, the approximate molecular weights of the isolated and solubilized apolipoproteins were characterized. Amino acid sequencing began, first with apoAII by Brewer et al. [23] and then later with other smaller peptides by others. Recently, the largest apolipoprotein (apo)B-100, considered to be the major atherogenic peptide, was solubilized as a monomer and this led to difficult but successful sequencing studies by several groups. These studies all document the early Donner and NIH concepts that the nature of the peptide is the fundamental determinant of the lipoprotein macromolecule.

Additionally, another landmark lipoprotein development was concerned with how various lipoproteins leave the bloodstream, and for what purpose. The novel approach of Goldstein and Brown [24] introduced the concept of lipoprotein cell receptors that led to the understanding of how LDL are normally bound at the cell surface and internalized by "receptors", and how the defective receptor in familial hypercholesterolemia leads to massive plasma LDL elevation and premature CHD. These cell "receptor" peptides also have now been isolated and sequenced.

Some of these developments, such as the final sequenching of the known apolipoproteins, were done simultaneously with the recent cell biology breakthroughs and developments. However, recently the genetic features of "lipoprotein diseases" are now being recognized, appreciated, and fully studied. But this concept was recognized earlier at NIH by the Fredrickson types of heritable lipoprotein patterns [25] as a basis for categorizing lipoprotein abnormalities, which are still used as clinical categories. Earlier, Gofman et al. [26] had described by AnUC the severe clinical types of *xanthoma tendinosum* (type II) and *xanthoma tubersum* (type III or dysbetalipoproteinemia).

Since about the mid-1970s, there has been an enormous expansion of the lipoprotein field. Many scientists have been crucially responsible for the development of lipoprotein studies and for identifying their vital role in the life process, and their contribution to the amelioration of premature coronary artery disease. Early pioneers were Svedberg

and Pickels, who provided the technology of the AnUC and the convenience of the preparative ultracentrifuge (and today, nearly every laboratory has one or more preparative machines). Lastly, Gofman was the conceptual pioneer, and it was his interpretations and early AnUC lipoprotein developments that helped build the "lipoprotein field" as we know it today. Figure 6 shows schematically the plasma lipoprotein classes as currently characterized by AnUC. This year over 1000 lipoprotein papers and abstracts from throughout the world have been published, in contrast with only one in the landmark year 1949.

Design and fabrication of Donner AnUC cells

Most of the early work done in analytic ultracentrifugation, (AnUC) involved sedimentation of macromolecules, however, the ultracentrifuge work done at Donner Lab of the Lawerence Berkeley Laboratory has emphasized the flotation of serum lipoproteins. Since the 1950s, the intense interest in the relationships between lipoproteins and atherosclerosis has created a demand for efficient and economical AnUC runs. Several modifications to the Beckman AnUC cells have been developed at Donner Lab to facilitate the thousands of flotation runs required by our research. The innovations to be described here include: 1) The manufacture of epoxy molds for centerpiece fabrication; 2) The fabrication of standard and 0.020 inch offset aluminum-filled epoxy centerpieces; 3) The inclusion of an overfill reservoir to equalize the meniscii in double-sectored centerpieces; 4) The use of bevelled sectorcups with wedge-window applications; and 5) Rotor-pinning and cell-slotting for simplified precise alignment of AnUC cells.

Since lipoprotein flotation by AnUC originates at the centerpiece's base-of-cell, and proceeds toward the meniscus, measurements of Schlieren patterns require a clearly defined base-of-cell. We simultaneously run a high- and a low-density serum preparation in two separate cells in order to visualize the full lipoprotein spectrum. To define unambiguous base-of-cell positions for the two runs, we use both a standard centerpiece (with its sectors centered in the centerpiece), and an offset centerpiece (whose sectors are set 0.020 inch (in.) radially inward.) To obtain this specification we find it most economical to manufacture all of our own centerpieces. The tooling and procedures for centerpiece fabrication

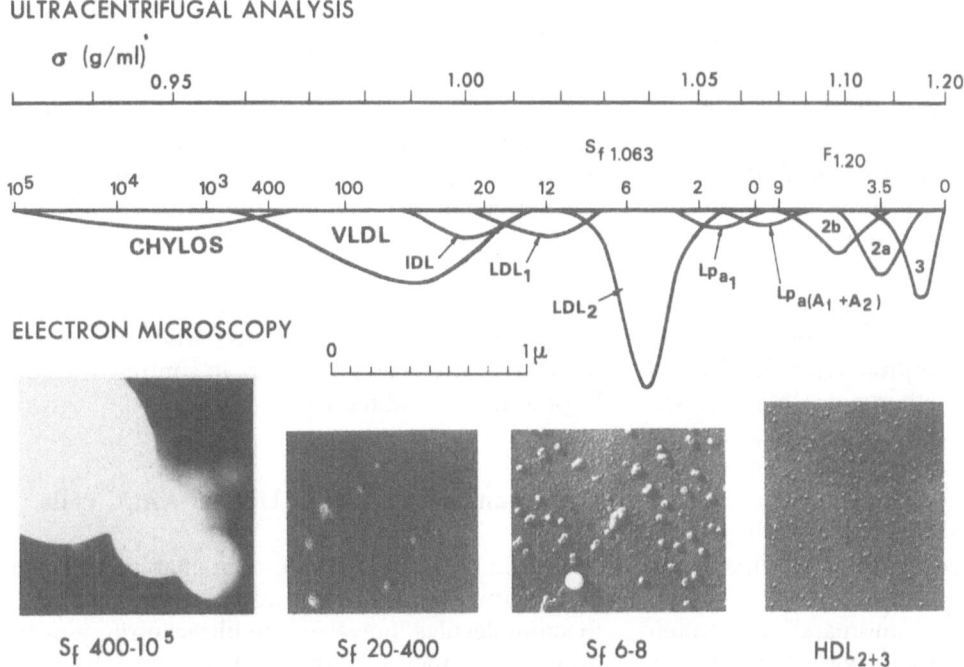

Fig. 6. Plasma lipoprotein classes as currently characterized by AnUC. Lp(a)-containing lipoproteins exist relatively low abundance and some are schematically shown at approximately 10-fold magnification. (Reprinted with permission from Ann NY Acad Sci (1980) 348:1—5.)

have been developed over the past 30 years at the Lawrence Berkeley Laboratory machine shops.

Epoxy mold manufacture

We machine centerpieces from aluminum-filled epoxy slugs cast in hardened tool-steel molds. The creation of standard and offset centerpieces requires two pairs of steel mold endpieces to position the sector cores within each type of mold. Figure 7 shows the mold housing and alignment key, the bottom end-piece supporting two highly polished sector cores, the top end-piece with vent holes, and a newly cast slug. The end-pieces, sector cores and key are made of hardened tool-steel which is ground precisely to size. The end-piece slots are produced by means of an electrical discharge machine, (EDM), which simplifies the manufacture of the end-pieces and increases the accuracy of the slot placement. First, holes are bored out to locate the sectors, next a carbon electrode is positioned in the holes and an electric current is applied to it, which erodes the metal for a "rough cut". Finally, a sectorshaped carbon electrode is inserted in the slot to make the final "cut", as pictured in Fig. 8.

Fig. 7. Centerpiece mold pieces and slug-pulling device

Sector cores are ground from hardened steel on a Sine Angle Plate until they are 0.002 inch oversized, then lapped on a glass lapping plate with a coarse compound, and polished on a Lapmaster plate fitted with 4/0 carborundum paper, until a mirror-like surface is obtained. The quality of this surface

Fig. 8. Electrical discharge machine set-up for final cut of sector slots

determines the quality of the finished centerpiece's sectors. The mold housing is made of unhardened steel with a keyway cut vertically by means of an EDM. A key, ground from hardened tool-steel and polished, positions the mold pieces precisely and forms the keyway in the final centerpieces.

Centerpiece fabrication

Before a mold is used it must be thoroughly coated with mold release 225 (Ram Chemicals, Div. of Whittaker Corp., Gardenia, California, USA). The large mold pieces may be coated with mold release by submersion, and then set in a 60°C oven to warm. The sector cores are polished lightly on the 4/0 carborundum-lined lapping plate, cleaned in acetone, and polished with lint-free tissues. A pipecleaner may be used to apply a thin uniform coat of mold release to the sector core surfaces. After warming for 5 min in a 60°C oven, the tacky surfaces may be gently polished with lint-free tissues and placed in the bottom end-piece sector slots. The mold pieces are assembled, except for the

top, and baked in a 107°C (225°F) oven for 30 min to set. The properly prepared mold should be equilibrated to 60°C before the epoxy mixture is poured.

After trying many different mixtures of Epon, aluminum, and curing agent, the most suitable recipe for centerpiece fabrication we have found is the following: Part A (Shell Epon 826) + Part B [(Ankamine K61B curing agent, equal to 10% of the Epon weight; E. V. Roberts & Associates, Culver City, California)] + Filler (Alcoa atomized aluminum #123, equal to 40% of the Epon weight). For a single casting, we normally mix 20 g Epon 826 + 8 g aluminum powder + 2 g curing agent. The mixture is quickly warmed to 75°C, placed in a vacuum chamber attached to a vacuum pump and deaerated under 25 inch vacuum until the frothy "head" collapses. The mixture is then poured into the prepared mold at 60°C. The mold top is put on and the whole mold is tapped gently with a mallet to settle the mixture. The poured mold is allowed to cure in a 60°C oven for at least 3 h but no more than 24 h and the hardened slug pulled from the mold with a slug pulling device, Fig. 7. The sector cores can be pressed out of the slug with a small "pusher" and a hand-operated press.

When 25—30 acceptable slugs have been cast, a lathe is set up with custom-made slug-holding fixtures for turning a slug to its finished diameter, Fig. 9. The standard and the offset slugholders are made to fit the taper of the chuck. The tailstock is fitted with a No. 2 Morse taper with a live center surrounded by an adjusting fixture. The chuck and tailstock fixtures each hold a set of four pins precisely set to fit snugly into the sectors of either a Standard or an Offset slug, depending upon which set of tools is used. By turning nylon screws in the adjusting fixture and indicating the slug-holding fixtures with a Last-Word gauge, both ends of the lathe can be aligned to within a few ten-thousandths of an inch. This degree of accuracy is necessary to avoid tapering of the slug. A tungsten-carbide tool bit is used to trim the slug diameter from 0.9 to 0.875 in. Slugs turned to diameter may be cut on a Do-All saw into two sections about 0.7 in. long.

The centerpiece thickness is next cut with the aid of a vacuum-chuck set-up (Fig. 10). The vacuum-chuck taper matches the taper of the spindle and is held by a vacuum-pipe which runs through the spindle to a rotating vacuum fitting and valve. Airways through the vacuum-chuck permit the center-

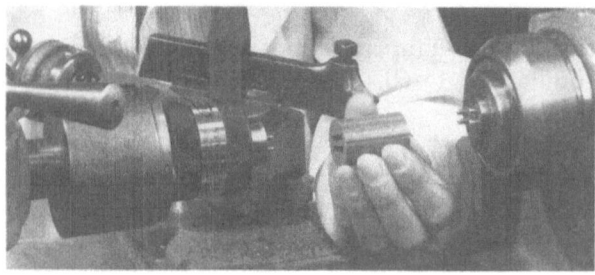

Fig. 9. Chuck and tailstock fixtures for turning slugs to a specific diameter

Fig. 10. Vacuum chuck set-up with centerpiece holding disc in place

piece-holding disc to be held firmly in place when a vacuum is applied. After one face is cut, the vacuum is released, the disc is reversed in the vacuum chuck, and an identical cut is made on the opposite face. The great advantage of this assembly is that it permits the centerpiece's faces to be cut uniformly parallel and at right-angles to its central axis.

A centerpiece is cut to a thickness of 0.480 in., just 0.004 in. over the finished thickness. Before a centerpiece is removed from the disc which holds it, a down-stepped ledge is cut from the perimeter of each side to reduce the face diameters to the diameter of a cell window. The final thickness, 0.475 in., is attained by a two-step process of hand lapping and polishing. Lapping is done on a 10 in. lapping plate with a coarse lapping compound and a

light-weight lapping oil. Each face is lapped in figure-eight patterns on four different regions of the plate for uniform wear. The centerpiece is turned a quarter-turn with each positional change to counteract the effects of a biased hand position. The polishing is done on a 6-in. plastic-polishing plate with a very fine grade of lapping compound and water. Normally, 0.0005 in. is removed from each face with polishing. The centerpiece faces should be within 0.0003 in. of parallel when finished.

The gasket hole and fill holes are drilled in each centerpiece with the aid of a special holding-form which has guide holes drilled in it (see Fig. 11). The tip of the gasket-hole end mill is ground to be concave so that the gasket holes will be slightly higher in the center where the fill holes emerge. The fillholes are drilled with a miniature precision drill-press fitted with a Nr. 68 drill. The holding fixture guides the drill at an angle from the gasket hole center. Only one set of gaskets and one screw-plug are needed for our double-sectored centerpieces. This adaptation helps to guard against cell-leaks since the screw-plug is radially aligned with the rotor.

The final step of our centerpiece fabrication process is the creation of an overfill reservoir and scribe-line. The same precision drill press used before is fitted with a Nr. 70 drill bit and a X—Y mechanical stage. A dial-gauge indicator, with

Fig. 11. Precision drill-press set-up for drilling overfill reservoir and drill-guides for fill-holes and crown-bore

1 inch of travel, is set to measure the changes in the Y direction. A dial indicator measuring in 0.0001-in. increments is set to read the vertical displacement of the drill chuck. The centerpiece, in a holding fixture, is positioned so that the hole will be drilled in the exact center of the septum and at the precise Y-displacement required by the cell. A hole is drilled slowly until half-way through the centerpiece, then the centerpiece is removed, turned over, and a matching hole is drilled through from the other side.

To scribe an over-flow channel 0.001 in. deep and 0.001 in. wide is a very delicate operation. We use a Nr. 20-gauge hypodermic needle which has been finely sharpened on the lapping plate. The needle is held stationary in the miniature drill-press chuck and a centerpiece is positioned precisely beneath it so that the needle is at the correct depth in one of the sector slots. By moving the mechanical stage in the X-direction, a scribe-line can be cut tangent to the top of the over-fill reservoir from sector to sector. The process is observed with a stereo-microscope set up beside the drill press; see Fig. 11.

We routinely fill the left sector of a centerpiece with a lipoprotein preparation and the right sector with a salt solution control whose density is equal to the lipoprotein fraction small molecule background. Aliquots of 0.420 mL are delivered to each sector with specially calibrated ground-glass tuberculin syringes. The meniscus of each filled sector is slightly above the position of the scribe-line. As a cell accelerates during a run, the excess volume in each sector is forced to drain into the over-fill reservoir through the tangential scribe-line. The reservoir volume is sufficient to contain the combined overfill volumes of both sectors. Thus, the meniscii in a double-sectored centerpiece are positioned at the same distance from their respective cell bases.

Before any centerpiece is used, it is soaked in a concentrated sodium hydroxide solution for 4—8 h to dissolve any aluminum particles exposed to its surface. This eliminates the possibility that the aluminum in the cell could react with the contents of a centerpiece. It also creates a pitted surface on the centerpiece faces which helps to form a tight seal between the windows and faces when the cell is torqued.

The complete Donner type AnUC X-cell is shown as an exploded diagram in Fig. 12. This shows a wedged sector-cup and 0° 49′ wedged quartz-crystal window combined with an offset centerpiece. The standard centerpiece is combined with a pair of plane quartz-crystal windows in two plane

Analytic Ultracentrifuge X-Cell Assembly
Donner Laboratory Design

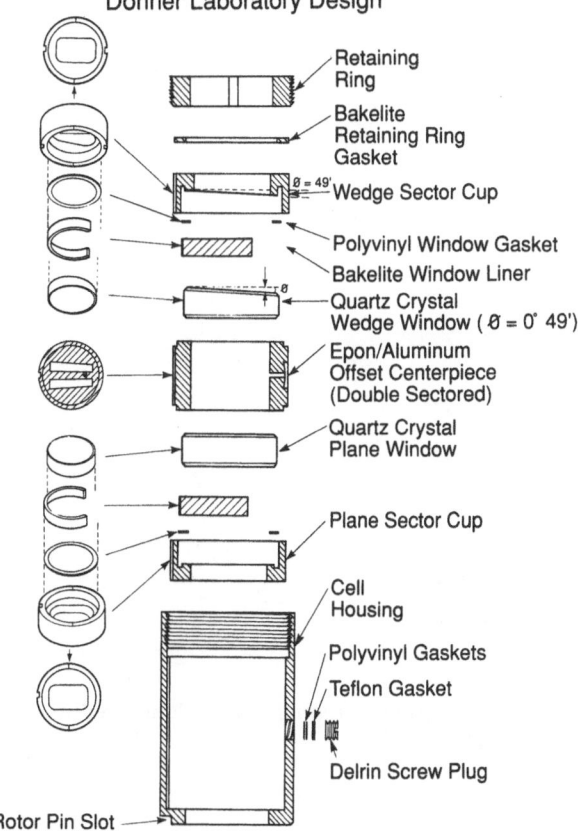

Fig. 12. Exploded diagram of Donner offset and wedge window combination AnUC cell. (Reprinted with permission from AOCS, Champaign, Illinois (1991): Orr JR et al., Perkins EG (Ed) In: Analyses of Fats, Oils, and Lipoproteins.)

sector cups. The rotor-pin slot is also shown at the base of the cell housing.

Figure 13 illustrates the relationship between the rotor pin and cell housing-slot. The pin is stationary on the centrifugal side of the rotor hole and precisely radially aligns a slotted cell housing when it is placed in the rotor hole. A specially designed reamer-guide assembly, pictured in Fig. 14, assures that the reamer will locate the pin holes exactly on the radius of the rotor.

Calibration of our Model E AnUC's are maintained by a special Beckman calibration cell (part No. 306386), as described in [27]. A calibration is performed immediately after the Model E Schlieren optical system is manipulated, cleaned, adjusted or whenever a drive is changed. Calibration factors on

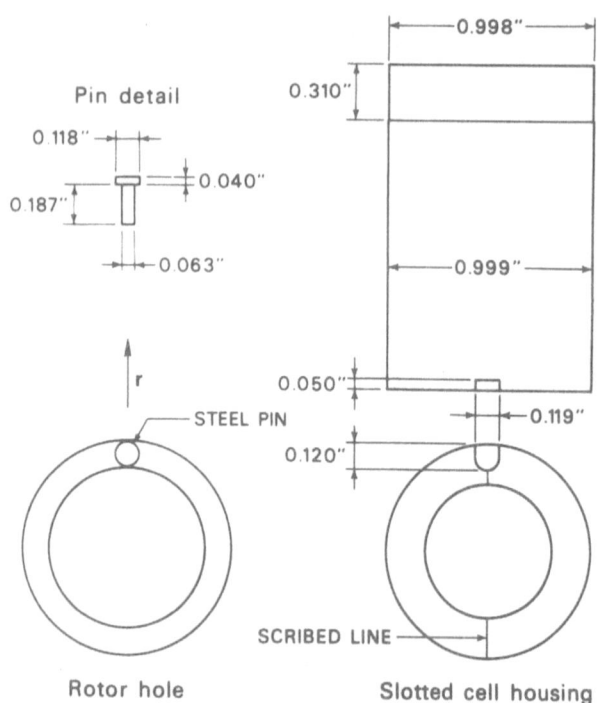

Fig. 13. Illustration of rotor pin and slotted cell housing specifications

Fig. 14. Rotor set-up with pin-hole reamer and reamer guides

all three of our Model E machines have been maintained to within ±0.5% for the past 24 years. During a 64-min flotation run at 52640 rpm, 10 photos are taken automatically at standard intervals to record the changing peak positions of the high- and low-density lipoprotein Schlieren patterns. Developed films are enlarged five times and projected onto paper templates calibrated specifically for the time intervals of greatest interest for lipoprotein resolution. Traced patterns are measured by a sonic digitizer connected to a VAX station 3200 computer. The sonic digitizer consists of two banks of microphones fixed to a table to form a right angle, and a button-operated sending unit with a "cross hair" window. A tracing is aligned and measured by placing the "cross hair" on a traced line and pressing the button at many points on each line. The (x, y) coordinates of each point are located sonically and transferred to a CRT screen which reproduces the traced patterns. A computer program subtracts the baseline control pattern from the lipoprotein/small molecule pattern, which leaves a Schlieren-profile of the lipoprotein alone. The area beneath the lipoprotein Schlieren curve is in-

tegrated and corrected for run temperature changes, Ogston-Johnston effects, flotation vs concentration effects and true sample density. The corrected areas are converted to mg/dL for each class of lipoprotein with the appropriate conversion factors.

The molecular weight of the principal low-density lipoprotein (LDL) component is calculated by means of a rho-intercept program. First the moving boundary flotation rate of LDL is calculated for both the high- and low-density runs. Next a plot of ηF vs. ρ is extrapolated to a flotation rate of zero, giving the particle density. See reference [28] for details.

Final results of our analysis include corrected Schlieren plots of both the low-density and high-density lipoproteins, a corrected flotation rate of the main species of LDL and HDL, quantitative data for each flotation interval measured in both the high- and low-density runs, and the molecular weight, particle diameter and particle density of the main LDL component.

Donner lab has been a pioneer in quantitative lipoprotein analysis by AnUC flotation and continues to perform state-of-the-art lipoprotein analysis, especially for collaborations. The role of AnUC in lipoprotein research has been decreased by both funding limitations and by developments in other non-quantitative methods of lipoprotein study, especially GGE. With renewed funding, now pending, we hope to continue as a resource to scientists involved in either lipoprotein or analytical ultracentrifuge methodology. Anyone desiring additional information, machine diagrams or clarification of any of the material presented is encouraged to contact the authors.

Acknowledgements

We thank Lynne Gloria and Penelope Henderson for manuscript preparation. This work was supported by NIH Program Project Grant HL 18574 from the National Heart, Lung, and Blood Institute of the National Institutes of Health, and was conducted at the Lawrence Berkeley Laboratory (Department of Energy contract DE-AC03-76SF00098 to the University of California).

References

1. Macheboef M (1929) Bull Soc Chem Biol 11:268
2. Tiselius A (1937) Trans Faraday Soc 33:524
3. Svedberg T, Rinde H (1924) J Am Chem Soc 46:2677
4. Svedberg T, Lysholm, A (1927) Nova Acta Regiae Soc Sci Upsaliensis, Vol Extra Ord
5. Lamm O (1933) Nature 132:820
6. McFarlane AS (1935) Biochem J 29:1175
7. Pederson KO (1947) Phys Colloid Chem 51:156
8. Oncley JL, Scatchard G, Brown A (1947) J Phys Colloid Chem 51:184
9. Pederson KO (1945) Ultracentrifugal studies on serum and serum fractions. Upsala 33
10. Pickels EG (1950) In: Uber FM (Ed) Biophysical Research Methods, New York, Interscience Publishers, Inc., pp 67—105
11. Johnston JP, Ogston AD (1946) Tr Farady Soc 42:789
12. Gofman JW, Lindgren FT, Elliott H (1949) J Biol Chem 179:973
13. Anitschkow N (1933) In: Cowdry EV (Ed) Arteriosclerosis, New York, Macmillan
14. Gofman JW, Lindgren F, Elliott H, Muntz W, Hewett J, Strisower G, Herring V, Lyon TP (1950) Science III:166
15. Shore B, Nichols AV, Freeman NK (1953) Proc Soc Exp Biol NY 83:216
16. Brown RK, Boyle E, Anfinson CB (1953) J Biol Chem 204:423
17. Havel RJ, Eder HA, Bragdon JH (1955) J Clin Invest 34:1345
18. Hillyard LA, Entenman C, Feinberg H, Chaikoff IL (1955) J Biol Chem 214:79
19. Lindgren FT, Nichols AV, Freeman NK (1955) J Phys Chem 59:930
20. Avigan J, Redfield R, Steinberg D (1956) Biochim Biophys Acta 20:557
21. Shore B (1957) Arch Biochem Biophys 71:1
22. Scanu A, Lewis LA, Bumpus FM (1958) Arch Biochem Biophys 74:390
23. Brewer HB, Lux SE, Ronan R, John KM (1972) Proc Natl Acad Sci USA 69:1304
24. Brown MS, Goldstein JL (1974) Science 185:61
25. Fredrickson DS, Levy RI, Lindgren FT (1968) J Clin Invest 47:2446
26. Gofman JW, De Lalla O, Glazier F, Freeman NK, Lindgren FT, Strisower B, Tamplin AR (1954) Plasma 2:413
27. Orr JR, Adamson GL, Lindgren FT (1991) In: Perkins EG (Ed) Analyses of Fats, Oils, and Lipoproteins, Chapter 28: Orr JR, Adamson GL, Lindgren FT (1991) Preparative Ultracentrifugation and Analytic Ultracentrifugation of Plasma Lipoproteins. AOCS, Champaign, Illinois
28. Lindgren FT, Jensen LC, Hatch FT (1972) The Isolation and Quantitative Analysis of Serum Lipoproteins. In: Nelson GJ (Ed) Blood Lipids and Lipoproteins. John Wiley-Interscience, New York, pp 181—274

Received June 10, 1991
accepted August 9, 1991

Authors' address:

J. R. Orr
Lawrence Berkeley Laboratory
University of California
Donner Laboratory
1 Cyclotron Road
Berkeley, California 94720, USA

Progress in Colloid & Polymer Science

Progr Colloid Polym Sci 86:12—22 (1991)

Analytical ultracentrifugation as a tool for studying membrane proteins*)

D. Schubert and P. Schuck

Institut für Biophysik der Johann Wolfgang Goethe-Universität, Frankfurt am Main, FRG

Abstract: Sedimentation equilibrium analysis in the analytical ultracentrifuge can be applied to membrane proteins solubilized by nonionic detergents. By representing a lipid-like environment for the proteins, the detergent micelles help to preserve the original tertiary and quaternary structure of the proteins and thus allow their native associations to be studied. The contributions of the membrane-bound detergent to the particle weight and the partial specific volume of the proteins or protein complexes can easily be taken into account, e.g., by the technique of density matching. The most important areas of application of the method are a) the study of reversibly self-associating systems, leading to the identification of the different oligomers present, and b) the analysis of stable or transient heterogeneous associations between membrane proteins. In the latter case, one of the proteins should be labeled with a dye in order to simplify the analysis. Sedimentation equilibrium experiments can also be applied to the analysis of the protein content of small lipid vesicles. Thus, analytical ultracentrifugation is apt to play an important role in studies on membrane proteins.

Key words: Sedimentation equilibrium; membrane proteins; molecular weight determination; self-association; heterogeneous associations

Introduction

Analytical ultracentrifugation is one of the classical methods in studying biological macromolecules [1, 2]. Most of its contributions in this field have, however, their origin in the period 1930s—1960s. During the last two decades, both the use of analytical ultracentrifugation and its reputation as a useful technique have decreased tremendously. There are two main reasons for this decline. The first one is connected with the introduction of SDS gel electrophoresis for the determination of the molecular weights of monomeric proteins and subunits of protein aggregates, which, until then, was — by sheer numbers — the most important application of analytical ultracentrifugation. The second reason is a technical one: the only instrument

available in most countries, the Spinco model E ultracentrifuge, has become obsolete, and the expense and difficulty in maintaining and operating it have continually increased. Curiously enough, the decline in the use of analytical ultracentrifugation has been paralleled by an enormous increase in its potential, due to the progress in computer sciences, so that this method now allows applications which were not feasible in the past. For many of these applications, analytical ultracentrifugation is unsurpassed by any other method. The availability of a newly developed analytical ultracentrifuge [3] should thus lead to a reanaissance of this technique.

The most important application of "modern" analytical ultracentrifugation certainly will be the study of associations between macromolecules, in particular the analysis of association-dissociation equilibria. The elucidation of both protein-protein and protein-nucleic acid interactions will benefit from this. A draw-back could be that, due to the

---

*) Dedicated to the pioneers in the field, Professors J. A. Reynolds and C. Tanford.

long period during which analytical ultracentrifugation was disregarded, the capabilities of the technique have become largely unknown to many workers in the field. This is particularly true for the area of biomembrane research, despite the importance of protein-protein interactions for both the structure and the function of biological membranes. The present review aims at improving this situation by describing how analytical ultracentrifugation could be used to analyze the associations between membrane proteins or between these proteins and those of the cytoplasm or the extracellular fluid. The emphasis will be on modern concepts to be applied rather than on a compilation of published data. The methodology described in the following can also be used with normal water-soluble proteins, simply by disregarding the complications which arise from the hydrophobicity of the membrane proteins.

Useful types of ultracentrifuge experiments

Most membrane proteins readily form aggregates. Frequently, the aggregation appears to be reversible. The behavior of reversibly associating systems during a sedimentation velocity experiment in the analytical ultracentrifuge will be governed, not only by structural, but also by kinetic factors. A reliable determination of the numerous parameters involved from the available experimental data does not seem to be feasible. On the other hand, the sedimentation equilibrium distributions of the systems will only be determined by structural parameters, namely the particle weight per mole M and the corresponding partial specific volume $\bar{v}$ of the aggregates present. In addition, it is, in the latter type of experiments, much easier to correct for the contributions of protein-bound detergent to the sedimentation behavior of the particles than in sedimentation velocity studies (see below). For these two reasons, sedimentation equilibrium experiments represent the method of choice when investigating membrane proteins. However, with most systems the number of parameters to be determined simultaneously will be large, even if the studies are performed by sedimentation equilibrium analysis. The experiments should thus be conducted in such a way that the description of the system can be simplified as much as possible. In this respect, the limitation to protein concentrations low enough to ensure ideal sedimentation behavior

seems to be obligatory. In addition, the use of a detection system which is able to focus on the protein-containing particles (in the presence of an excess of protein-free detergent micelles) or even on protein molecules selectively labeled with a dye may be necessary. The availability of a photoelectric scanner, in connection with a monochromator, thus seems to be a prerequisite. In the following, it will be assumed that the conditions mentioned will be fulfilled.

General considerations

Analytical ultracentrifugation can, of course, not be applied to the study of proteins in intact membranes, but requires solubilization and purification of the membrane proteins. One of the two classes of membrane proteins, the peripheral ones, can be isolated and handled in detergent-free solutions. They thus can be studied in the same way as normal water-soluble proteins. On the other hand, the members of the second class, the intrinsic (integral) membrane proteins are normally soluble in aqueous buffers only in the presence of detergent micelles. As discussed in more detail below, the protein conformation will be preserved in detergent micelles if suitable nonionic detergents are used [4, 5]. Simultaneously, the protein-detergent system is accessible to an analysis of protein particle or aggregate weight by analytical ultracentrifugation, as first shown by the group of Tanford and Reynolds [5—8]. It is this coincidence which makes analytical ultracentrifugation a most useful tool in studying associations between membrane proteins. On the other hand, it is the need to account for the bound detergent which, in ultracentrifuge studies, represents the peculiarity of membrane proteins as compared to water-soluble ones.

As shown very recently, membrane proteins reconstituted in small lipid vesicles represent another model system which can be analyzed by analytical ultracentrifugation. It can be used for studies on the relationship of the state of association of a transport protein and its function, by combining ultracentrifuge and transport measurements [9].

Influence of detergents on protein conformation

Ionic detergents, in particular SDS, will tend to denature the proteins and will bind both to polar and nonpolar regions of the protein. If, on the other hand, suitable nonionic detergents at concentrations higher than their critical micelle concentration (CMC) are used, the detergent micelles will incorporate the protein without changing its conformation by perfectly mimicking the protein's environment in the intact membrane: the hydrophobic belt of the membrane protein being located in the hydrophobic interior of a detergent micelle and the protein's more polar regions again being surrounded by water or covered by the polar headgroups of the detergent molecules [4, 5]. As a consequence of the preservation of protein conformation in the detergent micelles, the proteins will also tend to show the same types of associations as in the intact membrane. The protein-detergent system thus represents a suitable model for studying protein-protein associations involving membrane proteins.

It should be noted that nonionic detergents of the polyether type (e.g., the members of the Triton or Brij series) are unstable and that their degradation products may oxidize membrane proteins [10, 11]. This, in turn, may change the state of association of the proteins [12, 13]. A thorough search for a time-dependence of the state of association may be necessary to detect such effects.

Allowing for protein-bound detergent or lipid [5—8]

The quantity which characterizes a particle's sedimentation equilibrium distribution is the term $M(1 - \bar{v}\rho)$ (ρ: density of the solvent). For a protein embedded in a detergent micelle, the bound detergent as well as lipid molecules, eventually bound to the protein, will contribute to the term, which now will characterize the protein-detergent-lipid complex. We thus may rename it $M_C(1 - \bar{v}_C\rho)$, where the subscript C refers to the complex. The quantity of primary interest will, of course, remain to be the corresponding term for the pure protein, $M_P(1 - \bar{v}_P\rho)$. The relationship between the terms is given by Eq. (1):

$$M_C(1 - \bar{v}_C\rho) = M_P(1 - \bar{v}_P\rho) + M_D(1 - \bar{v}_D\rho)$$

$$+ M_L(1 - \bar{v}_L\rho) , \qquad (1)$$

where M_D and M_L are the particle weights of the detergent or lipid, respectively, per mole of complex and $\bar{v}_D$ and $\bar{v}_L$ are the corresponding partial

specific volumes[1]). For a discussion of the thermodynamic basis of Eq. (1) see [5, 7, 8].

When studying protein/detergent complexes, M_L is usually either zero or $\ll M_P$, and $\bar{v}_L$ close to unity [5]. The last term of Eq. (1) thus can be disregarded in most cases, or $M_C(1 - \bar{v}_C\rho)$ can be corrected for it, by means of the lipid content determined from phosphorus analysis.

The contribution of the detergent can easily be taken into account [5—8]. Its contribution to $M_C(1 - \bar{v}_C\rho)$ can be calculated with sufficient accuracy, if M_D can be measured, e.g., by gel filtration or equilibrium dialysis experiments using radioactively labelled detergent [14]. If M_D is unknown, the contribution of the detergent can be blanked out by varying the solvent density. For uniform protein-detergent complexes, a procedure applicable to all types of detergents in use consists of measuring $M_C(1 - \bar{v}_C\rho)$ at different solvent densities, by substituting H_2O by appropriate D_2O/H_2O mixtures, and extrapolating the data to $\rho = 1/\bar{v}_D$. At that point, $1 - \bar{v}_D\rho$ vanishes, so that $M_P(1 - \bar{v}_P\rho)$ can be determined without knowledge of M_D. By combining data obtained at different solvent densities, it is also possible to determine the amount of protein-bound detergent (provided that $\bar{v}_P$ and $\bar{v}_L$ are known, see below) [12, 15]. A variant of the procedure, applicable to detergents with $\bar{v}_D$-values between 0.9 and 1.0 ml/g, consists of performing the experiments just at the solvent density $\rho = 1/\bar{v}_D$. This condition can be realized for a large number of detergents commonly in use [5]. The D_2O/H_2O ratio, necessary for matching the detergent density, can be determined with great accuracy from sedimentation equilibrium runs on the protein-free detergent, by varying the D_2O content until no sedimentation of the detergent micelles will occur [16]. With detergents transparent in the wavelength range accessible to the photoelectric scanner, a small amount of highly absorbing apolar or amphiphilic molecules of a density near to that of the detergent (e.g., phenol) may be added to allow monitoring the concentration distribution (D. Schubert, unpublished data). In all cases, H-D exchange will lead to small (and often negligible) changes in M_P

---

[1]) Tanford and Reynolds prefer to use an alternative form of the expression $M_C(1 - \bar{v}_C\rho)$, namely $M_P(1 - \phi\rho)$, where ϕ is the volume increment per g of protein, measured under conditions where the chemical potentials of all solvent components are kept constant [5, 7, 8].

and $\bar{v}_P$, for which corrections may be applied if the amino acid composition of the protein is known [7].

An alternative procedure is to use a detergent for which $\bar{v}_D$ is very close to 1.00 ml/g, so that its contributions to $M_C(1 - \bar{v}_C\rho)$ in aqueous solutions are negligible [17]. Of course, with the protein under study, the use of a special detergent may be disadvantageous for other reasons [18].

In studying the self-association or the mixed associations of membrane proteins, it may not be necessary to blank out the contribution of the detergent to $M_C(1 - \bar{v}_C\rho)$ if the corresponding terms for the protein protomers can be determined. The mixed associations of erythrocyte band 3 protein with ankyrin or hemoglobin have been analyzed, e.g., without density matching, by applying the individual $M_C(1 - \bar{v}_C\rho)$-values obtained from experiments on the isolated proteins [19, 20] (see below).

Molecular weight of uniform protein particles

Until two decades ago, the determination of polypeptide molecular weights, including those of membrane proteins, represented the main application of analytical ultracentrifugation. This can be done reliably by sedimentation equilibrium analysis, provided that conditions can be found for completely monomerizing the protein [5]. Nevertheless, this application is only of minor importance at present since polypeptide molecular weights are now routinely determined either from sequence data (obtained from cDNA sequencing) or from SDS gel electrophoresis. With respect to the latter technique, however, it is clear that, with some membrane proteins, this method may yield figures which differ from the true values by a factor up to three. A well-known example is glycophorin A, the major glycoprotein of the erythrocyte membrane [5, 14]. On the other hand, analytical ultracentrifugation is the best technique for the determination of the particle molecular weight and, thus, of the state of association of membrane proteins in solutions of nonionic detergents. As discussed above, the results elucidate the state of association of the proteins in the intact membrane.

The basis of all particle weight determinations by sedimentation equilibrium analysis in the analytical ultracentrifuge is represented by Eqs. (2) and (3): The equilibrium distribution of uniform particles is given by

$$c(r) = c(r_0)\exp[M(1 - \bar{v}\rho) \cdot A(r, r_0, \omega)] , \qquad (2)$$

where

$$A(r, r_0, \omega) = \omega^2(r^2 - r_0^2)/2RT ;$$

($c(r)$, $c(r_0)$ are particle concentration at a distance r and a fixed distance r_0, respectively, from the rotation axis; ω is the angular velocity of the rotor. This is equivalent to the well-known equation

$$\frac{d\ln c}{dr^2} = \frac{M(1 - \bar{v}\rho)\omega^2}{2RT} . \qquad (3)$$

Application of Eq. (3) by plotting $\ln C$ vs r^2 and determining $M(1 - \bar{v}\rho)$ from the slope of the straight lines obtained is one of the classical methods of measuring the molecular weight of monomeric proteins and of uniform protein aggregates. When $M(1 - \bar{v}\rho)$ is substituted by $M_C(1 - \bar{v}_C\rho)$, Eqs. (2) and (3) are also valid for protein-detergent complexes. Equation (3) will thus yield $M_P(1 - \bar{v}_P\rho)$, if one of the procedures for accounting for protein-bound detergent is applied (see above). Self-association of the proteins will lead to nonlinear $\ln C(r^2)$-plots. In this case, the local slope of the curves will yield, instead of M, the weight average molecular weight, M_W [1, 2].

Values for $\bar{v}_P$, required for the determination of M_P from the experimental $M_P(1 - \bar{v}_P\rho)$-data, can be calculated from the composition of the proteins [8]. The "true" $\bar{v}_P$ may differ from the calculated one by up to approx. ±0.02 ml/g [16, 21].

Applications of the principles described above to intrinsic membrane proteins can be found, e.g., in [7, 12–14, 22, 23]. (Useful advice for performing the measurements is given in [8].)

Analysis of self-associating proteins

Many, if not most membrane proteins apparently exist in the form of different oligomeric aggregates [5, 24, 25]. Frequently, the associations seem to be reversible. As discussed above, they apparently persist after solubilization of the proteins by nonionic detergents [4, 5].

Studies on mixtures of oligomeric aggregates of a single polypeptide chain (or of a stable protein complex) represent an important application of analytical ultracentrifugation. With reversibly aggregating systems, this method is by far the best one available. The most simple and useful approach is based on an extension of Eq. (2): For a mixture of different oligomers of a protein protomer of molecular weight M_1 and complex molecular weight M_{1C} (including bound detergent and lipid), the concentration distribution $c(r)$ in the ultracentrifuge cell, at sedimentation equilibrium, will be the sum of the contributions of the different oligomers present:

$$c(r) = \sum_i c_i(r)$$

$$= \sum_i c_i(r_0) \exp\left[iM_{1C}(1-\bar{v}_{iC}\rho) \cdot A(r, r_0, \omega)\right] , (4)$$

where the subscript i denotes the number of protein protomers per oligomer (see, e.g., [26, 27]). Assuming that the partial specific volumes of the different oligomers are identical (which may not always be the case; see below), $\bar{v}_{iC}$ may be substituted by $\bar{v}_C$. For reversibly self-associating systems, Eq. (4) may also be written in terms of the monomer concentration c_1 and the equilibrium constants K_{1i} [26, 27].

In the analysis according to Eq. (4), sums of Boltzmann distributions, which represent the different oligomers expected, are fitted to the experimental data by least squares procedures. If M_{1C} and $\bar{v}_C$ are known or if $M_{1C}(1 - \bar{v}_C\rho)$ can be determined experimentally (or, under the conditions of density matching, $M_{1P}(1 - \bar{v}_P\rho)$), a system of linear equations for the unknown parameters $c_i(r_0)$ has to be solved. A straightforward algebraic solution exists in this case [26]. If $M_{1C}(1 - \bar{v}_C\rho)$ is unknown (or if nonidealities have to be taken into account), nonlinear regression will have to be applied, which may be complicated by highly correlated parameters [28—31]. Equations with less correlation can be derived under the assumption that the protein mass in the solution is conserved during the run (which may not always be the case) [29]. According to our experience, it is highly recommendable to determine $M_C(1 - \bar{v}_C\rho)$ in separate experiments, also because the experimental $c(r)$-data rarely allow a reliable simultaneous determination of more than 3—4 unknown parameters. We and others [27] have also observed that, for a reliable discrimination between different models of self-association, the $c(r)$-data set used should be as large as possible and should contain both flat and steep parts.

A successful fit to the experimental data, based on a special model of self-association, will yield the local concentrations of the different oligomers considered in the calculations. An example concerning erythrocyte band 3 protein is shown in Fig. 1. Of course, other models of self-association will have to be ruled out by separate calculations and subsequent comparison of the quality of the fits. The approximate total amount of each oligomer in the cell may be obtained by integrating Eq. (4) between the meniscus and the cell bottom, using the calculated figure for $c_i(r_0)$ [9]. The data do not show, however, whether the different oligomers are stable or are linked in an association equilibrium. To answer this question it may be utilized that, with reversibly associating systems, $M_W(c)$-data obtained under different conditions of initial protein concentration or rotor speed should be identical [2, 26, 27].

An alternative approach for analyzing self-association is derived from methods for the evaluation of light scattering and small angle x-ray scattering data. In the first step, the weight average molecular weight $M_W(c)$ is determined from $\ln c(r^2)$-plots according to Eq. (3). Afterwards, equation based on reasonable models of self-association are fitted to the $M_W(c)$-data [32, 33]. As discussed elsewhere [30], this approach seems to be inferior to that discussed above.

It is possible, with both approaches described, to combine the information obtained at different initial protein concentrations or different rotor speeds. This is obvious when the fitting procedures are based on M_W-values. In evaluations according to Eq. (4), averaged $c(r)$-data sets may be used, if necessary after appropriate weighting or after correction for the differences in rotor speed [28, 31].

In the evaluation procedures discussed, it is normally assumed that all oligomers present have the same partial specific volume (including bound detergent). It has been claimed that this assumption may be grossly incorrect, both for Apo-A-I from human high-density lipoprotein (in detergent-free buffers) [34] and for myelin proteolipid protein (in solutions of Triton X-100) [35]. In the case of Apo-A-I the finding is, however, not supported by the results of other groups [36, 37]. Concerning the data shown in [35], it seems to us that a more probable interpretation is that the samples studied were strongly heterogeneous with respect to the state of association.

The study of mixed associations

Mixed associations between membrane proteins seem to occur even more frequently than self-

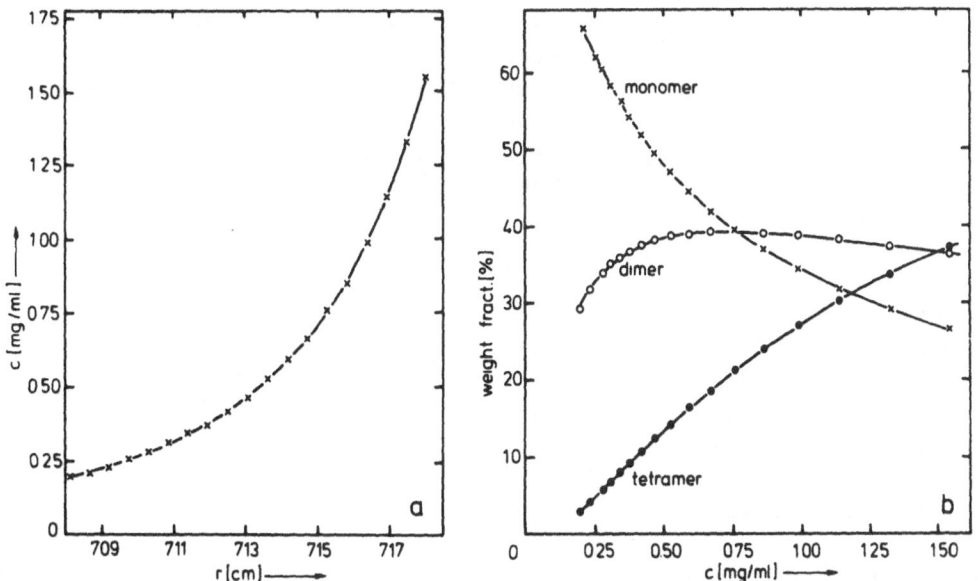

Fig. 1. Analysis of the self-association of erythrocyte band 3 protein in Triton X-100. The solvent density was adjusted to match that of the detergent. a) Experimental concentration-vs-radius data, obtained at a rotor speed of 14000 rpm (×), with a least squares fit to the data based on a monomer-dimer-tetramer model of self-association according to Eq. (4) (———). b) Relative contributions of the different oligomers to local protein concentration. Supplementary data show that the oligomers are in an association equilibrium [13].

association (see, e.g., [38, 39]). To analyze them is a difficult task. At least in the case of unstable aggregates, sedimentation equilibrium analysis of the proteins in solutions of nonionic detergents seems by far to be the best approach for studying details of the associations, e.g., their stoichiometries [19, 20].

Two proteins, a and b, which form mixed aggregates of stoichiometry 1:1, but do not show self-association, will lead to a sedimentation equilibrium distribution $c(r)$ given by Eq. (5):

$$c(r) = c_a(r) + c_b(r) + c_{ab}(r) , \qquad (5a)$$

where the distributions for the uncomplexed proteins, $c_a(r)$ and $c_b(r)$, will have the form indicated by Eq. (3) and the contribution of the complex $c_{ab}(r)$ is given by

$$c_{ab}(r) = c_{ab}(r_0) \exp[(M_{aC}(1 - \bar{v}_{aC}\rho)$$
$$+ M_{bC}(1 - \bar{v}_{bC}\rho)) \cdot A(r, r_0, \omega)] , \qquad (5b)$$

(which represents a generalization of Eq. (4)).

If we introduce a quantity $\bar{v}_C^*$ (e.g., identical either to $\bar{v}_{aC}$ or $\bar{v}_{bC}$) and corresponding quantities M_{aC}^* and M_{bC}^* defined by Eqs. (5c) and (5d):

$$M_{aC}(1 - \bar{v}_{aC}\rho) = M_{aC}^*(1 - \bar{v}_C^*\rho) \qquad (5c)$$

$$M_{bC}(1 - \bar{v}_{bC}\rho) = M_{bC}^*(1 - \bar{v}_C^*\rho) , \qquad (5d)$$

we may rewrite Eq. (5b) as

$$c_{ab}(r) = c_{ab}(r_0) \exp[(M_{aC}^* + M_{bC}^*)(1 - \bar{v}_C^*\rho)$$
$$\cdot A(r, r_0, \omega)] . \qquad (5e)$$

In the general case where both protein a and b show self-association and, in addition, form mixed aggregates of different stoichiometries, the resulting concentration distribution $c(r)$ will be given by Eq. (6):

$$c(r) = \sum_i c_{ai}(r) + \sum_j c_{bj}(r) + \sum_i \sum_j c_{aibj}(r) , \qquad (6a)$$

with

$$c_{aibj} = c_{aibj}(r_0) \exp[(iM^*_{1aC} + jM^*_{1bC})(1 - \bar{v}^*_C \rho)$$

$$\cdot A(r, r_0, \omega)] \, . \tag{6b}$$

It is clear that, in the general case described by Eq. (6), a reliable fit to the experimental $c(r)$-data is not feasible and that the equation will be useful only if the number of unknown parameters used is low. This implies i) that the number of different aggregates formed by the proteins is relatively small; ii) that it is possible to focus on the particles of special interest, i.e., the mixed aggregates; and iii) that, possibly, several terms of Eq. (6a) may be combined, at least in an initial stage of the computations.

A general advice can be given with respect to item ii): it can be best satisfied by labeling one of the proteins with a dye [19, 40, 41]. By monitoring the sedimentation equilibrium profiles at a wavelength where only the dye will absorb, only contributions of the labeled protein and of its complexes will be detected. This will eliminate one of the sums in Eq. (6b) and, thus, greatly reduce the number of parameters to be determined from the fit, especially if the labeled protein shows a relatively simple self-association. Of course, it has to be made sure that labeling the protein does not change its association behavior.

An illustration of the procedure described is given by recent studies on the association of the erythrocyte membrane proteins band 3 and ankyrin [19, 42]. This association links the membrane skeleton to the lipid bilayer; it plays an important role in the structure and stability of the erythrocyte membrane [38, 39]. In solutions of nonionic detergents, the band 3 protein is in a monomer/dimer/tetramer association equilibrium [12, 13] (which also seems to be the state of association of the protein in the erythrocyte membrane [43—45]), whereas ankyrin is monomeric [38]. The mixed aggregates formed by the two proteins were identified by sedimentation equilibrium analysis according to Eq. (6), after labeling the ankyrin with fluorescein isothiocyanate and by making use of the reduction in the number of terms allowed by this approach. Two striking results were obtained from this analysis: a) The absorbance-vs-radius distributions (measured at 495 nm) could be perfectly fitted by assuming that the only dye-labelled particles present in the samples were free ankyrin and a com-

plex of one ankyrin molecule and four band 3 molecules. b) The contributions to total absorbance of complexes containing one or two band 3 molecules could be shown to be negligible. Thus, it is only the band 3 tetramer (and not the monomer or dimer) which serves as ankyrin binding site (see Fig. 2).

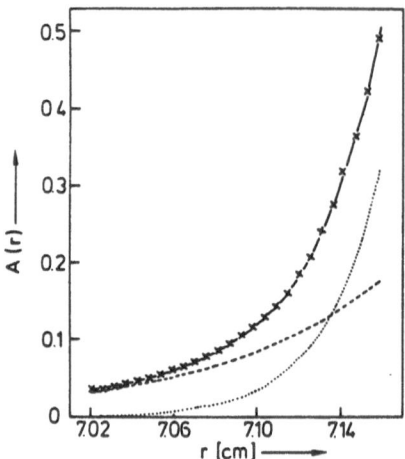

Fig. 2. Experimental absorbance-vs-radius distribution, $A(r)$, measured at a wavelength of 495 nm for a mixture of dye-labeled ankyrin and erythrocyte band 3 protein in solutions of the nonionic detergent nonaethylene glycol lauryl ether ($\times$), and the theoretical curve fitted to the data assuming that the only dye-labeled particles present in the solution were free ankyrin and a complex of one ankyrin molecule and the band 3 tetramer (———). The figure also shows the calculated contributions to $A(r)$ of free ankyrin (– – –) and the complex (...). The values of $M_{1C}(1 - \bar{v}_C \rho)$ for the two proteins were determined in separate experiments [19]

The importance of item iii) in the analysis of heterogeneous associations is demonstrated by a recent study on the associations of erythrocyte band 3 protein with the oxygenated form of hemoglobin (oxy-Hb) [20]. In analogy to the case described above concerning the band 3-ankyrin association, the study aimed at identifying those band 3 oligomers which are able to bind oxy-Hb. In addition, it aimed at determining the stoichiometries of the complexes. The band 3-Hb system offers the advantage that the heme group represents an intrinsic dye label, so that there is no need for chemically modifying one of the proteins and subsequently testing for the preservation of its association behavior. On the other hand, it is complicated, not only by the presence of five homopolymers (monomers, dimers, and tetramers of band 3, dimers and tetramers of oxy-Hb), but also by the large number of different heterogeneous aggregates expected: according to data obtained with band 3 fragments, each band 3 molecules has binding sites for two dimers or one tetramer of oxy-Hb [39, 46]. However, an inspection of the possible molecular weights of the heme-containing complexes reveals that their range can be divided into three subranges which are clearly separated from each other and are characterized by the band 3 content of the particles. It, thus, was possible to use a two-step analysis. In a first step it was assumed that each band 3 oligomer capable of binding Hb will bind one Hb tetramer. By this assumption, the double sum in Eq. (6a) is replaced by a single one, and the number of parameters to be fitted is reduced to 5 when the fitting procedure is based on absorbance-vs-radius data obtained at 505 nm. The evaluation showed that it is only the band 3 tetramer which acts as an oxy-Hb binding site. In the second step of the analysis, the influence of the assumed molecular weight of the band 3-Hb complex, M^*, on the sum σ of the squared residuals (the error sum) in the fit was studied. It turned out that the $\sigma(M^*)$-curves showed a well-defined minimum, but that the position of this minimum on the M^*-axis depended on the concentrations of band 3 and Hb used in the experiments: the minimum M^*-values found correspond to an aggregate containing, per band 3 tetramer, one or two Hb dimers, whereas the upper limit corresponds to aggregates containing 4 Hb tetramers per band 3 tetramer (Fig. 3).

It is clear that, in studies of the type described in this section, $M_{1C}(1 - \bar{v}_C\rho)$-values determined in separate experiments should be used in order to reduce the number of variables.

The methods outlined above could help to elucidate the associations of a large number of different membrane proteins.

Determination of the protein content of lipid vesicle membranes

The molecular weight or the state of association of proteins embedded in lipid vesicle membranes can, in principle, be determined in the same way as described for proteins embedded in detergent micelles: In the case of homogeneous protein-lipid vesicles, $M_C(1 - \bar{v}_C\rho)$ can be obtained according to

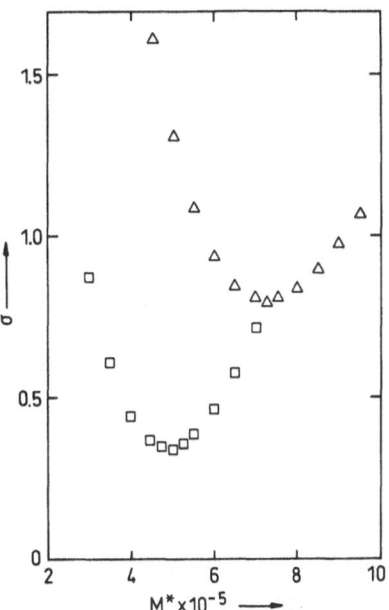

Fig. 3. Fits to the absorbance-vs-radius distributions at 505 nm of band 3-hemoglobin mixtures: dependency of the sum σ of the squared residuals on the assumed molecular weight M^* of the band 3-hemoglobin complex, for a model in which the band 3 tetramer represents the only hemoglobin binding site. The two curves belong to experiments performed at different concentrations of the two proteins and correspond to the maximum and minimum values of M^* obtained [20]

Eq. (3), and the use of density matching allows the direct determination of $M_P(1 - \bar{v}_P\rho)$ applying Eq. (1). In the latter case, application of Eq. (4) will also allow to analyze the distribution of "protein molecules per vesicle" in heterogeneous vesicle populations. In contrast to the situation with protein-detergent complexes, the molecular weight of the lipid part of the complexes is, however, much larger than that of the protein (e.g., phosphatidylcholine vesicles of diameter 150 nm have a molecular weight of approx. $1.5 \cdot 10^8$ [47]). Any deviation from ideal density matching thus will lead to a large error in the apparent value of $M_P(1 - \bar{v}_P\rho)$. Under most conditions, this error will completely obscure the exact protein content of the vesicles [9]. Two requirements have to be met in order to allow a sufficiently precise determination: the precision of the density matching has to be as high and the diameter of the vesicles (and thus M_L) has to be made as small as possible. In a recent study on erythrocyte band 3 protein — the anion transport protein — these requirements have, for the first time, been met by the following procedure [9]: 1) The proteolipid vesicles of average diameter 70 nm, as obtained by reconstituting the anion transport system in egg phosphatidylcholine membranes, were transformed into smaller ones by treatment in a French press. From the vesicle mixture, unilamellar vesicles of diameter (32 ± 3) nm were isolated by gel filtration. 2) Protein-free 32 nm-vesicles prepared in the same way as the corresponding proteoliposomes, in solutions of different D_2O content, were studied in equilibrium sedimentation experiments in the analytical ultracentrifuge (monitoring the concentration distribution by measuring the apparent absorption due to light scattering), and the H_2O/D_2O ratio was varied until no sedimentation of the vesicles was observed. The D_2O content initially determined could be reproduced within ±0.5 mg/ml if D_2O addition was done by weight. The optimum D_2O content, however, had to be redetermined when a different lipid batch or vesicles of different diameter or differently prepared were used[2]. Subsequent ultracentrifuge studies on the proteoliposomes clearly demonstrated that the 32 nm-vesicles contained a single band 3 molecule when prepared by the use of unmodified band 3 protein, whereas vesicles which were prepared using crosslinked band 3 dimers contained a single copy of the dimers (Fig. 4). Measurements of anion efflux from the vesicles, performed parallelly, showed that the number of anions transported per band 3 molecule and minute was identical in both cases. This proves that the band 3 monomer can transport anions and that dimerization of the protein has little or no influence on the protein's transport capacity [9].

It is obvious that the methodology described above should be applicable to other membrane-bound transport proteins for which the functional oligomeric state is unknown.

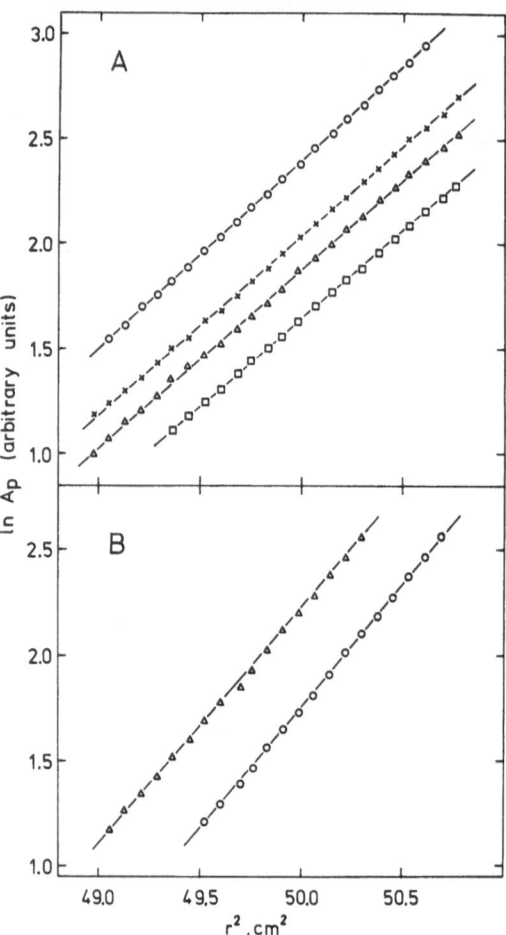

Fig. 4. Analysis of the content of erythrocyte band 3 protein of "small" lipid vesicles, under the condition of density matching: Plots of the logarithm of the apparent absorbance A_p of the protein-containing vesicles vs r^2, from sedimentation equilibrium runs on four different vesicle samples containing unmodified band 3 protein (A), and on two batches containing crosslinked band 3 dimers (B). Rotor speed: 12000 rpm (A) and 10000 rpm (B). The resulting M_p-values are 100500 ± 2500 (A) and 198500 ± 2500, and correspond to those of a band 3 monomer and dimer, respectively [9]

---

[2]) In addition, the incorporation of a protein molecule into the vesicle membrane may influence the partial specific volume of that part of the lipid which forms the immediate surrounding of the protein (the "boundary" or "annular" lipid [48]). However, if the $\bar{v}_L$-value of 30 lipid molecules surrounding a protein of $M_p = 100000$ [48] would be changed by ±0.02 ml/g, the error in $M_p(1 - \bar{v}_p\rho)$ which results from ignoring the local change in $\bar{v}_L$ would only be ±2%.

Conclusions

It is clear from the foregoing that analytical ultracentrifugation is still a powerful and indispensible tool in biochemistry, despite the decline in both the use and the reputation of the technique during the last two decades. What has happened during that time is that the main area of application of the method has changed from the determination of molecular weights of homogeneous particles to the study of complex associations between macro-molecules, in particular of association-dissociation equilibria and thus of unstable aggregates. For these applications, analytical ultracentrifugation is superior to all other techniques available. Sedimentation equilibrium runs, combined with mathematical analyses of the concentration distributions by least squares techniques, represent by far the most useful type of studies in modern analytical ultracentrifugation. Membrane proteins solubilized by nonionic, nondenaturing detergents constitute a field of special importance. In these detergent solutions, homogeneous as well as heterogeneous associations between membrane proteins or between membrane proteins and soluble ones can be studied in a lipid-like environment. The complications brought about by the contributions of protein-bound detergent to the molecular weight and the partial specific volume of the proteins can be easily taken into account, e.g., by the technique of density matching. Sedimentation equilibrium analysis, combined with density matching, can also be used for the determination of the protein content of small lipid vesicles. When applied to transport proteins reconstituted in lipid bilayer membranes, this allows to study the relationship between a protein's oligomeric state and its transport function. There can be little doubt that, with the availability of a newly developed analytical ultracentrifuge, increasing use will be made of the potential offered by this technique.

Acknowledgements

The original work from the authors' group discussed in this review was performed in collaboration with and with the financial support by the Max-Planck Institute for Biophysics, Department of Cell Physiology, Frankfurt (Director: Prof. Dr. H. Passow). Additional financial support was provided by the Deutsche Forschungsgemeinschaft (SFB 169) and from the Stiftung Volkswagenwerk (project 11-2004).

References

1. Schachman HK (1959) Ultracentrifugation in Biochemistry. Academic Press, New York
2. Fujita H (1975) Foundations of Ultracentrifugal Analysis. John Wiley & Sons, New York
3. Schachmann HK (1989) Nature 341:259—260
4. Helenius A, Simons K (1975) Biochim Biophys Acta 415:29—79
5. Tanford C, Reynolds JA (1976) Biochim Biophys Acta 457:133—177
6. Tanford C, Nozaki Y, Reynolds JA, Makino S (1973) Biochemistry 13:2369—2376
7. Reynolds JA, Tanford C (1976) Proc Natl Acad Sci USA 73:4467—4470
8. Reynolds JA, McCaslin DR (1985) Methods Enzymol 117:41—53
9. Lindenthal S, Schubert D (1991) Proc Natl Acad Sci USA 88:6540—6544
10. Lever M (1977) Anal Biochem 83:274—281
11. Chang HW, Bock E (1980) Anal Biochem 104:112—117
12. Pappert G, Schubert D (1983) Biochim Biophys Acta 730:32—40
13. Schubert D, Boss K, Dorst HJ, Flossdorf J, Pappert G (1983) FEBS Lett 163:81—84
14. Grefrath SP, Reynolds JA (1974) Proc Natl Acad Sci USA 71:3913—3916
15. Edelstein SJ, Schachman HK (1967) J Biol Chem 242:306—311
16. Schubert D, Boss K (1985) Z Naturforsch 40c:908—911
17. Ludwig B, Grabo M, Gregor J, Lustig A, Regenass M, Rosenbusch JP (1982) J Biol Chem 257:5576—5578
18. Butler PJG, Kühlbrandt W (1988) Proc Natl Acad Sci USA 85:3797—3801
19. Mulzer K, Kampmann L, Petrasch P, Schubert D (1990) Colloid Polym Sci 268:60—64
20. Schuck P, Schubert D (1991) FEBS Lett (submitted)
21. Durchschlag H, Jaenicke R (1982) Biochem Biophys Res Commun 108:1074—1079
22. Reynolds JA, Stoeckenius W (1977) Proc Natl Acad Sci USA 74:2803—2804
23. Reynolds JA, Karlin A (1978) Biochemistry 17:2035—2038
24. Tanford C (1977) In: Abrahamsson S, Pascher J (eds) Structure of Biological Membranes. Plenum, New York, pp 497—508
25. Schubert D (1988) Molec Aspects Med 10:233—237
26. Haschemeyer RH, Bowers WF (1970) Biochemistry 9:435—445
27. Osborne JC, Bronzert TJ, Brewer HB (1977) J Biol Chem 252:5756—5760
28. Hensley P, O'Keefe CD, Spangler CJ, Osborne JC, Vogel CW (1986) J Biol Chem 261:11038—11044
29. Lollar P (1987) Biophys Chem 28:245—251
30. Milthorpe BK, Jeffrey PD, Nichol LW (1975) Biophys Chem 3:169—176
31. Johnson ML, Correia JJ, Yphantis DA, Halvorson HR (1981) Biophys J 36:575—588
32. Adams ET, Williams JW (1964) J Am Chem Soc 86:3454—3461
33. Tang LH, Powell DR, Escott BM, Adams ET (1977) Biophys Chem 7:121—139
34. Formisano S, Brewer HB, Osborne JC (1978) J Biol Chem 253:354—360
35. Gow A, Winzor DJ, Smith R (1985) Biochim Biophys Acta 828:383—386
36. Stone WL, Reynolds JA (1976) J Biol Chem 250:8045—8048

37. Vitello LB, Scanu AM (1976) J Biol Chem 251:1131—1136
38. Bennett V (1983) In: Elson E, Frazier W, Glaser L (eds) Cell Membranes: Methods and Reviews. Vol II, Plenum, New York, pp 149—195
39. Low PS (1986) Biochim Biophys Acta 864:145—167
40. Osborne JC, Powell GM, Brewer HB (1980) Biochim Biophys Acta 619:559—571
41. Servillo L, Brewer HB, Osborne JC (1981) Biophys Chem 13:29—38
42. Mulzer K, Petrasch P, Kampmann L, Schubert D (1989) Stud Biophys 134:17—22
43. Dorst HJ, Schubert D (1979) Hoppe-Seyler's Z Physiol Chem 360:1605—1618
44. Jennings ML (1984) J Membrane Biol 80:105—117
45. Passow H (1986) Rev Physiol Biochem Pharmacol 103:61—203
46. Cassoly R (1983) J Biol Chem 258:3859—3864
47. Mimms LT, Zampighi G, Nozaki Y, Tanford C, Reynolds JA (1981) Biochemistry 20:833—840
48. Gennis RB (1989) Biomembranes: Molecular Structure and Function. Springer, New York, pp 191—215

Received June 10, 1991
accepted August 9, 1991

Authors' address:

Prof. Dr. D. Schubert
Institut für Biophysik der J. W. Goethe-Universität
Theodor-Stern-Kai 7, Haus 74
6000 Frankfurt/M. 70, FRG

Progress in Colloid & Polymer Science   Progr Colloid Polym Sci 86:23—29 (1991)

Can sedimentation analysis contribute to the protein folding problem?

R. Jaenicke and K. Lehle

Institut für Biophysik und Physikalische Biochemie, Universität Regensburg, Regensburg, FRG

Abstract: Protein folding is a spontaneous reaction commonly modeled by in vitro reconstitution experiments. Ultracentifugal analysis may be applied to characterize the initial, intermediate, and final states in the denaturation-renaturation cycle. With respect to intermediate states, the "molten globule state" and the "A-state" are considered to be of major importance. Using α-lactalbumin, lactate dehydrogenase, γII-crystallin and immunoglobulin as examples, compact intermediates observed at low pH may be analyzed by ultracentrifugation. Correcting for changes in the partial specific volume, and considering aggregation as a side reaction, previous conclusions regarding the increase in hydrodynamic radius of α-lactalbumin at pH 2 are questionable. In the case of lactate dehydrogenase, subunit dissociation and charge effects on the partial specific volume are superimposed in the sedimentation characteristics at low pH. s-values and reactivation kinetics prove that the enzyme in its "A-state" consists of "structured monomers" with significant residual structure. γII-crystallin is exceptional in that it shows long-term stability even at pH 1. Addition of chaotropic agents, e.g., urea leads to sequential unfolding of its two domains. Immunoglobulin at low pH forms an "alternatively folded state" with residual secondary *and* tertiary structure. Cooperative reversible thermal transitions prove this state to be qualitatively different from the "molten globule". Ultracentrifugal analysis allows the spectral and calorimetric data to be interpreted in structural terms.

Key words: Protein folding; ultracentrifugation; gamma-crystallin; gamma-globulin; lactate dehydrogenase; molten globule

Introduction: The folding problem

The question of "how do proteins gain their native three-dimensional structure?" is presently one of the most challenging problems in physical biochemistry and biotechnology. Evidently, if we knew the answer in terms of an algorithm relating a given amino acid sequence to its unique spatial configuration, designing proteins with desired biological functions would be feasible. So far, this "second half of the genetic code" has not been solved [1]. One reason is its complexity: in contrast to Nirenberg and Khorana's genetic code at the DNA level, the protein folding code cannot be colinear in the sense that specific short stretches of the amino acid sequence could be directly correlated with local structural elements in the context of the final structure of the complete molecule [2, 3]. A second reason is based on the observation that the code must be highly degenerate: within narrow limits of root mean square deviations, one and the same topology is found to be coded by a wide variety of different sequences. For example, over the whole time span of evolution, the hemoglobins have preserved only two out of ca. 150 amino acid residues without significantly altering the overall topology of the molecule. Since the folding process within the cell is most difficult to assess, denaturation-renaturation experiments have been the common approach to analyze protein folding. Since Anson and Mirsky's [4] and Anfinsen's [5] classical work they have been considered clear proof for the idea that protein folding means "self-organization", i.e., spontaneous and autonomous structure formation without the requirement of extrinsic factors, nor the input of energy. Recently, this view has been challenged by the finding that the rate-limiting steps in the overall process are catalyzed by specific enzymes and "chaperones" or "polypeptide chain binding proteins" [6—9]. However important "foldases" may turn out to be for the process of in vivo folding, they mainly refer to the *kinetics* of the reactions involved without affecting the *equilibrium* properties. These are characterized by the general scheme

$$N \rightleftharpoons I_i \rightleftharpoons U ,$$ (1)

which indicates that folding/unfolding occurs as an ordered sequential reaction with a series of intermediates (I_i) on the pathway from the unfolded (U) to the folded, native (N) state or vice versa.

In vitro experiments mostly refer to the (essentially irreversible) denaturation-renaturation process

$$N \rightarrow U_i \rightarrow N^* ,$$ (2)

where N and N^* stand for the native and renatured states, and U_i for a whole variety of possible unfolded states, depending on the denaturing conditions. The kinetics of the folding processes in vivo must be fast compared to the generation time of an organism, i.e., they must occur within seconds to minutes. Obviously, this is a time range that a priori excludes ultracentrifugal analysis as a means to follow the process. What can be done is to characterize the initial and finals tates (N, U_i, N^*) with regard to their hydrodynamic properties and their quaternary structure; in addition, knowing that solvent parameters (temperature, pH, specific solvent components, etc.) are crucial for the kinetics of the reaction, variations of the experimental conditions may allow stable or metastable intermediates to be characterized. The outcome of previous studies in this connection may be summarized as follows: i) the amino acid sequence and the solvent determine the native three-dimensional structure of a given protein; ii) in most cases, refolding after preceding denaturation leads back to the native state ($N^* = N$); iii) intermediates on the folding pathway can be populated under specific solvent conditions so that they become accessible to detailed experimentation, including sedimentation analysis [10, 11].

Native, denatured, and renatured states

The characterization of proteins in their native state is established since Svedberg's early ex-

periments [12]. In order to compare the N and N^* states, the analysis should include the search for by-products, especially aggregates, using meniscus depletion [13] or boundary analysis [14].

The "denatured protein" (U_i) represents an ensemble of readily interconvertible conformers with clearly similar energies. Under ideal solvent conditions, the polypeptide chain will approach the random coil. It is presently still unclear to what extent an unfolded polypeptide chain conforms to this model. Since there is no "good solvent" for the backbone *and* the chemically diverse side-chains of all 20 amino acids, even in strong denaturants some non-random behavior is expected to remain, not to mention the space-requirements of the real chain which may trap short-range interactions [9]. Experimental evidence proves that the extent of unfolding of globular proteins differs widely for different denaturants, from local distortions of the polypeptide backbone or loosening of domains to "randomization". Spectral data clearly indicate that even in concentrated guanidinium chloride, proteins still retain some residual structure [15].

In order to characterize the product of renaturation, by-products have to be removed by gel-filtration or centrifugation [9—11]. Heeding this precaution, it has become clear that reconstituted proteins not only exhibit full catalytic activity and unchanged K_M values, but also identical spectral and hydrodynamic properties (Table 1). The major by-products are high molecular weight aggregates which represent irregular networks with a broad molecular weight distribution of highly structured particles, possibly in a state comparable to the "molten globule" (see below). As one would predict for a polymerization reaction involving partially folded monomeric units, renaturation and aggregation compete with each other such that, at high protein concentrations, aggregation becomes the predominant reaction [16].

The "molten globule state"

The *equilibrium* unfolding of proteins commonly obeys the two-state model

$$N \rightleftharpoons U$$ (3)

with only the folded native and the unfolded states populated [17]. Obviously, this approach does not yield much information with respect to the pathway

Table 1. Comparison of the native, denatured and renatured states of γII-crystallin from calf eye lens and porcine muscle lactate dehydrogenase (LDH-M$_4$)[a]

	γII-Crystallin			Lactate dehydrogenase		
	N	U_U	N^*	N	U_A	N^*
$s^0_{20,w}$ (S)	2.40	1.60	2.42	7.60	1.9	7.56
M (kDa)	19.8	20.0	20.0	140	36.6	142
A_{sp} (IU/mg)	—	—	—	640	0	655
λ_{max} (nm)	320	350	320	339	336	339
F_{rel} (%)	100	118	102	100	34	104
$-\Theta_{222nm} \cdot 10^{-3}$	4.0	0.9	4.0	15.4	11.5	16.0

[a] For preparation of γII-crystallin, cf. [11]; LDH-M$_4$ from Boehringer Mannheim. N, N^*: native and renatured states; U_U, denatured state in 8 M urea after 5 h incubation in 0.1 M NaCl/HCl pH 2.0 (20°C); U_A, denatured state in 1 M glycine/H$_3$PO$_4$ pH 2.3 after 5 min incubation (0°C). Renaturation by dilution as reported [2, 10, 11].

$s^0_{20,w}$, sedimentation constant (corrected for water viscosity and 20°C, extrapolated to zero protein concentration); M, weight-average molecular mass from sedimentation velocity and high-speed sedimentation equilibrium runs: Beckman Spinco Model E with high sensitivity UV-scanner, AnD and AnG rotors and double sector cells. Partial specific volume determinations: DMA 10 (A. Paar, Graz). A_{sp}, specific activity; λ_{max}, F_{rel}: maximum of fluorescence emission (λ_{exc} = 280 nm) and relative maximum fluorescence intensity: Hitachi F 4000 (λ_{exc} = 280 nm). Θ_{222nm}, molar ellipticity in the far-UV (degr · cm^2 · dmol^{-1}): JASCO J-600A spectropolarimeter. For details cf. [10].

Table 2. Characteristics of the "molten globule state" of proteins [20, 21]

Method	Property
Circular dichroism, infrared spectra	Native-like secondary structure
Nuclear magnetic resonance, near UV-circular dichroism, absorbance, and fluorescence	Denaturation-like alteration of the environment of aromatic residues
Solvent perturbation spectroscopy	Change in exposure of tryptophan, unaltered titration of tyrosin
Ultacentrifugation	Increase in Stokes radius, aggregation caused by exposed hydrophobic surface
Differential scanning calorimetry	Non-cooperative thermal transition
Polarization of tryptophan fluorescence	Slow intramolecular structural fluctuations

of structure formation. In contrast, *kinetic* evidence from spectroscopic and other techniques allows a whole series of intermediates to be analyzed, suggesting that acquisition of the native three-dimensional structure proceeds as a "hierachical condensation reaction": Short stretches of secondary structure first combine to local supersecondary structures which subsequently either merge to native-like intermediates, or collapse to form the "molten globule state" [18—21]. This is defined as "the preferred conformational state of the unfolded protein *under refolding conditions*" [19]. Characteristics of this metastable set of structural intermediates are summarized in Table 2. The most important property in the present context is the "expansion" of the protein molecule, compared with the native state which was first reported for α-lactalbumin [20, 22, 23]. The reference state used to characterize the "molten globule" as structural intermediate is remarkable insofar as in previous work the protein at pH 7.0—8.6 and at pH 2.0 was compared. The respective data (s^0_{20} = 1.93 S and 1.6 S with a concomitant increase in the hydrodynamic radius from 19 to 23, assuming 20—30% hydration [23], on one hand, and 1.67 S and 1.40 S

[20], on the other) do not consider the effect of pH on the partial specific volume [12]. This has been shown to be significant [24] so that the incomplete Svedberg correction [12] leads to an erroneous result. Considering the range of error in the extrapolation of s_{20} to zero protein concentration [23], the sedimentation data used as a reference for the "molten globule state" are inappropriate. Redetermining the sedimentation constants for α-lactalbumin (Sigma, Deisenhofen) in 50 mM sodium phosphate buffer pH 7.0—8.5 and 2.0 yields $s_{20,w}^0$ = 1.84 ± 0.03 S and 1.48 ± 0.04 S, respectively [25]. Since no additional information with respect to changes in hydration and/or anisotropy is available, no attempt is made to interpret these figures. In spite of the high electrostatic potential at pH 2, α-lactalbumin is found to form aggregates apart from the compact "2S particle". The enhanced binding of anilinonaphthalene-sulfonic acid points to an increase in hydrophobic surface area which might accompany the expansion of the molecule, giving rise to aggregation.

Acid-induced unfolding and structure formation

In the acid pH range, a wide variety of proteins is known to possess residual structure which may even be promoted at pH < 2. The corresponding "A-state" has been claimed to show certain characteristics of the "molten globule state". However, depending on the anion, the native tertiary structure may be preserved to a significant extent [2, 10, 11, 26, 27].

According to the previous definition, α-lactalbumin *at pH 2* may be assumed to represent the "molten globule" as a *trapped intermediate* accessible to experimentation due to charge-induced stabilization. In the following, three examples will be given where proteins exhibit anomalous behavior at acid pH without fulfilling the previously mentioned criteria for either the "A-state" or the "molten globule state" (Table 2).

γII-crystallin from calf eye lens is a highly stable protein which remains in its native conformation during the whole life-span of the organism. Its stability refers to temperature (≤ 75 °C), urea (up to 7 M), and pH (pH 1—10). The protein consists of two homologous domains, each composed of two similar "Greek key" motifs. Upon unfolding/refolding (in urea at low pH), a reversible biphasic transition is observed which shows that the two do-

mains unfold and fold independently [28]. The corresponding three-state equilibrium transition

$$N \rightleftharpoons I \rightleftharpoons U \tag{4}$$

has been confirmed by kinetic experiments and limited proteolysis [28, 29]. As shown in Fig. 1, the sedimentation analysis of the system in the presence of increasing urea concentrations confirms the profile obtained from fluorescence emission measurements at 360 nm, pH 2.0. Altering the pH from 1 to 7 does not affect the sedimentation constant. Obviously, the native all-β structure is preserved over the whole pH range, without the requirement of specific anions. Thus, γII-crystallin at low pH shows neither A-state behavior nor the characteristic properties attributed to the "molten globule state". This conclusion is corroborated by independent evidence gained from spectroscopic and calorimetric experiments [28].

Lactate dehydrogenase from pig muscle (LDH-M$_4$) is a homotetramer with a subunit molecular mass of 36 kDa. Its structure and stability, as well as its folding/association properties have been in-

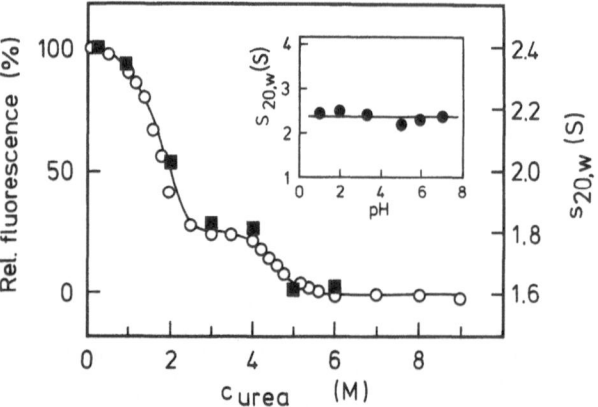

Fig. 1. Urea-induced unfolding of γII-crystallin in 0.1 M NaCl/HCl pH 2.0 at 20 °C. Relative fluorescence emission at 360 nm (λ_{exc} = 280 nm), 0.04 mg/ml protein concentration (○), and sedimentation coefficient $s_{20,w}^c$ (0.2 mg/ml protein concentration) (■), after 24 h incubation at given urea concentrations. Fluorescence emission of the native state and the fully denatured state are taken as 100% and 0%. Sedimentation velocity experiments were performed in an AnG rotor using the native protein as a reference: 40000 rpm, 12 mm double sector cells. Data corrected for the effect of urea and pH, using bovine serumalbumin as standard [24]. Insert: pH-dependence of $s_{20,w}^c$ (S): 1—24 h incubation in 0.1 M Teorell-Stenhagen buffer. 0.2 mg/ml protein concentration; sedimentation conditions as indicated

vestigated in detail [2, 30, 31]. At pH values close to the pK of aspartate and glutamate, the tretramer undergoes dissociation to the monomer, due to the repulsion of the protonated carboxylate groups (Fig. 2A). Decreasing the pH further leads to an increase of $s_{20,solv}$ which levels off when the pH effect on the partial specific volume is taken into account [24]. The resulting limiting s value indicates a compact "collapsed" state of the molecule. This is obvious if the sedimentation coefficients at low pH in the presence and in the absence of 6 M guanidinium chloride are compared; the latter is found to exceed the corrected value in 6 M guanidine, pH 2, by a factor of ≈ 2.3.

As taken from reactivation experiments starting from the two different unfolded states (M_A and M_G, unfolded by acid and guanidine, respectively), it becomes clear that only in 6 M guanidinium chloride is the protein fully unfolded: the reconstitution shows the typical sigmoidal folding/association profile (Fig. 2B, cf. [2]). At pH 2, the profile is hyperbolic, showing that reactivation in this case starts from the "structured monomer", i.e., folding steps involved in the process must be fast compared to the mixing time, again in accordance with the "molten globule" concept. Fitting the kinetics by the simplest possible mechanism, the sigmoidal reactivation profile turns out to be uni-bimolecular (requiring one first-order and one second-order rate constant, k_1 and k_2, only)

$$4\,M_G \xrightarrow{k_1} 4\,M \xrightarrow{fast} 2\,M_2 \xrightarrow{k_2} M_4 \,. \quad (5)$$

On the other hand, the hyperbolic curve can be described by only one rate-determining second-order rate constant (k_2); it represents a simple bimolecular reaction:

$$4\,M_A \xrightarrow{fast} 4\,M \xrightarrow{fast} 2\,M_2 \xrightarrow{k_2} M_4 \,. \quad (6)$$

In both mechanisms, the final product M_4, is indistinguishable from the native enzyme (Table 1).

Immunoglobulin (IgG) is a 150 kDa glycoprotein of two heavy and two light chains linked together by disulfide bonds. The all-β structure of the domains, together with the covalent cystine bridges provide a high stability of the protein in a wide pH range from 3 to 10 [32]. At pH < 3, an "alternatively folded state" is observed which differs from all the other states discussed previously (cf. [20, 21, 26, 27]), as it shows a high degree of secondary structure, increased hydrophobicity, a native-like fluorescence

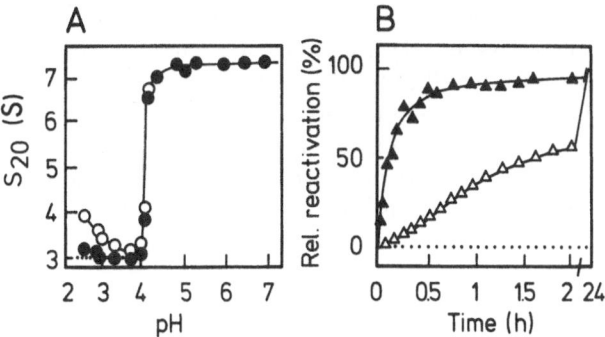

Fig. 2. pH-dependent dissociation of porcine muscle lactate dehydrogenase (LDH-M$_4$), 0.42 mg/ml in 0.1 M phosphate in the presence of 1 mM EDTA and 0.1 mM dithiothreitol, 20°C.

A: Effect of pH on the sedimentation coefficient: (○) uncorrected $s^c_{20,solv}$ data calculated from sedimentation velocity runs at 40000 rpm (data taken from [31]); (●) $s^c_{20,w}$, i.e., the same data after correcting for the pH dependence of the partial specific volume, using bovine serumalbumin as standard [24].

B: Kinetics of reconstitution after denaturation in 6 M guanidinium chloride (△) and 0.1 M H$_3$PO$_4$ pH 2 in the presence of 1 M (NH$_4$)$_2$SO$_4$ (▲), respectively ($c_{LDH} = 5$ µg/ml). Sigmoidal profile calculated according to Eq. (5), with $k_1 = 8 \times 10^{-4}$ s^{-1} and $k_2 = 3 \times 10^4$ M$^{-1} \cdot$ s^{-1}. Hyperbolic profile calculated according to Eq. (6) with a second-order rate constant $k_2 = 2.4 \cdot 10^4$ M$^{-1} \cdot$ s^{-1}. In fitting the kinetic data, monomers and dimers were assumed to be catalytically inactive; reactivation parallels tetramer formation. For further details, cf. [10]

emission maximum, and a highly cooperative, reversible thermal transition, indicative of well-defined secondary structure *and* tertiary contacts [32]. The cooperative equilibrium transition (which is generally absent in the case of the "molten globule") is different from the irreversible transitions observed at higher pH values, thus proving that the "alternatively folded state" represents a distinct stable conformation different from both the N state and the "molten globule state".

As taken from ultracentrifugal analysis (Table 3), immunoglobulin in its "alternatively folded state" retains its compact structure with a sedimentation coefficient only slightly below the characteristic value for the "7S particle", but far beyond the value determined for the unfolded protein in 6 M guanidinium chloride. The respective data have been corrected for the charge effect on the partial specific volume; densimetric determinations of the latter at 20°C yielded 0.719 ± 0.002 cm$^3 \cdot$ g^{-1} at pH 2.0 and 0.735 ± 0.002 cm^3 · g^{-1} at pH 7.0. As one would expect, the molecular mass remains constant under all conditions summarized in Table 3. In the "alternatively folded state", the protein shows a tendency to aggregate. However, the process is sufficiently slow as not to disturb the analysis, even under conditions of high-speed sedimentation equilibrium. The sedimentation coefficients at acid and alkaline pH signal similar alterations in the hydrodynamic properties. This does not necessarily mean that there is an equivalent "state" at alkaline pH; spectral data seem to contradict such "symmetrical behavior" [32].

Table 3. Ultracentrifugal analysis of immunoglobulin (MAK 33) at varying pH[a])

pH	$s^c_{20,solv}$ (S)	$s^c_{20,w}$ (S)	M (kDa)
2.0	6.03 ± 0.02	5.53 ± 0.02	164.2 ± 6.0
2.7	6.14 ± 0.02	5.72 ± 0.02	
4.2	6.60 ± 0.09	6.49 ± 0.09	
7.0	6.60 ± 0.04	6.60 ± 0.04[b])	153.8 ± 3.4
10.2	6.45 ± 0.08	6.20 ± 0.08	
11.9	6.12 ± 0.04	5.58 ± 0.04	
2.0[c])	1.7 ± 0.2	4.2 ± 0.3	146 ± 10

[a]) Sedimentation analysis in 40 mM potassium phosphate at 20°C, protein concentration 0.67 mg/ml: Beckman Spinco Model E, AnG rotor, double sector cells with sapphire windows. Sedimentation coefficients were calculated from log$r$ vs t plots and corrected according to [12]. M values from high-speed sedimentation equilibrium runs at 12000 rpm. For further details, cf. [25].
[b]) For pH 7.0, extrapolation of $s^c_{20,w}$ to zero protein concentration yields $s^0_{20,w}$ = 6.93 S.
[c]) In 6 M guanidinium chloride; partial specific volume corrected for solvent conditions, using bovine serumalbumin as standard [24].

Conclusions

The acquisition of the native state of globular proteins *in vivo* is commonly simulated by *in vitro* reconstitution experiments. These start from the denatured polypeptide chain where "denatured" stands for any local or global non-native conformation of the protein. As has been discussed in this paper, ultracentrifugal analysis has useful applications in the elucidation of protein folding only as far as (meta-)stable states are concerned. These refer, first, to the native and renatured states. Sedimentation coefficients, particle weights and particle distributions have been shown to be most useful in proving or disproving the identity of the initial and final states in a denaturation-renaturation cycle. Second, they refer to structural intermediates trapped under specific solvent conditions. An example illustrating this possibility has been the characterization of the acid denatured state of α-lactalbumin in terms of the "molten globule" model. In this connection, it has been shown that corrections with respect to alterations of the solvent, including effects on the partial specific volume, may be decisive. In addition, one has to keep in mind that the old wisdom that alterations in sedimentation rates may reflect changes in mass, shape and/or solvation (hydration) also holds for the "molten globule" and any intermediate that can be detected by hydrodynamic methods. Sedimentation equilibrium measurements have to be included to dissect at least part of the contributions. In order to obtain an estimate with respect to shape and solvation, independent methods such as (elastic) lightscattering, neutron scattering (with or without contrast variation), or buoyancy measurements have to be included.

Regarding the outcome of sedimentation data in connection with the protein folding problem, there has been clear proof that under optimum conditions the product of reconstitution is indistinguishable from the native starting material. Since high molecular weight aggregates are the main by-product in the course of the reaction, ultracentrifugation is the method of choice, not only to quantify N^*, but also to determine the amount and size of aggregates. There are no experimental methods available which could be applied to analyze the denatured states of proteins in a satisfactory way. All that ultracentrifugal analysis is able to show is that different denaturing solvent conditions lead to a wide variety of sedimentation characteristics. The broad distribution of readily interconvertible conformations does not allow a detailed description. This holds for partially unfolded states, as well as the "fully unfolded" protein. Available data are still fragmentary and, in part, controversial [2, 19, 33—35]. The question is significant since the folding mechanism may be determined by residual structural elements or local structures preserved even in strong denaturants [9].

High-resolution nuclear magnetic resonance may be applied to resolve this problem [36].

Acknowledgements

Work was supported by grants of the Deutsche Forschungsgemeinschaft and the Fonds der Chemischen Industrie. R. J. wishes to thank the Fogarty International Center for Advanced Study, NIH, Bethesda, for generous support and hospitality.

References

1. Gierasch L, King J (eds) (1990) Protein folding: Deciphering the second half of the genetic code. American Assoc Advancement of Science
2. Jaenicke R (1987) Progr Biophys Mol Biol 49:117—237
3. Jaenicke R (1988) In: Huber R, Winnacker E-L (eds) Protein structure and protein engineering. 39. Colloquium Mosbach. Springer Verlag, Berlin, Heidelberg, New York, pp 16—36
4. Anson ML (1945) Adv Prot Chem 2:361—384
5. Anfinsen CB (1973) Science 181:223—230
6. Rothman JE (1989) Cell 59:591—601
7. Ellis RJ (1990) Seminars Cell Biology 1:1—9
8. Fischer G, Schmid FX (1990) Biochemistry 29:2205—2212
9. Jaenicke R (1991) Biochemistry 30:3147—3161
10. a) Jaenicke R, Rudolph R (1986) Meth Enzymol 131:218—250
    b) Jaenicke R, Rudolph R (1989) In: Creighton TE (ed) Protein structure: A practical approach. IRL Press, Oxford, New York, pp 191—223
11. Siebendritt R (1989) Dissertation, Universität Regensburg
12. Svedberg T, Pedersen KO (1940) Die Ultrazentrifuge, Verlag T. Steinkopff, Dresden, Leipzig, p 32
13. Yphantis DA (1964) Biochemistry 3:297—310
14. Baldwin RL, van Holde KE (1960) Fortschr Hochpolym Forsch 1:451—511
15. Baldwin RL (1989) Trends Biochem Sci 14:291—294
16. Zettlmeissl G, Rudolph R, Jaenicke R (1979) Biochemistry 18:5567—5571
17. Privalov PL (1970) Adv Prot Chem 33:167—241
18. Gò N (1984) Adv Biophys 18:149—164
19. Creighton TE (1990) Biochem J 18:149—164
20. Dolgikh DA, Abaturov LV, Bolotina IA, Brashnikov EV, Bychkova VE, Bushnev VN, Gilmanshin RI, Lebedev YO, Semisotnov GV, Tiktopulo EI, Ptitsyn OB (1985) Eur Biophys J 13:109—121
21. Kuwajima K (1989) Proteins: Struct Funct Genet 6:87—103
22. Kronman MJ, Holmes LG, Robbins FM (1967) Biochim Biophys Acta 133:46—55
23. Kronman MJ, Andreotti RE (1964) Biochemistry 3:1145—1160
24. Durchschlag H, Jaenicke R (1982) Biochem Biophys Res Comm 108:1074—1079
25. Lehle K (1991) Diplomarbeit Universität Regensburg
26. Goto Y, Calciano LJ, Fink AL (1990) Proc Natl Acad Sci USA 87:573—577
27. Goto Y, Takahashi N, Fink AL (1990) Biochemistry 29:3480—3488
28. Rudolph R, Siebendritt R, Neßlauer G, Sharma AK, Jaenicke R (1990) Proc Natl Acad Sci USA 87:4625—4629
29. Sharma AK, Minke-Gogl V, Gohl P, Siebendritt R, Jaenicke R, Rudolph R (1990) Eur J Biochem 194:603—609
30. Holbrook JJ, Liljas A, Steindel SJ, Rossmann MG (1975) In: Boyer PD (ed) The Enzymes 3rd ed, Vol 11. Academic Press, New York, pp 191—292
31. Rudolph R, Jaenicke R (1976) Eur J Biochem 63:409—417
32. Buchner J, Renner M, Lilie H, Hinz H-J, Jaenicke R, Kiefhaber T, Rudolph R (1991) Biochemistry 30:6922—6929
33. Damaschun G, Damaschun H, Gast K, Gernat C, Zirwer D (1991) Biochim Biophys Acta, in press
34. Damaschun G, Damaschun H, Gast K, Zirwer D, Bychkova VE (1991) Int J Biol Macromol, in press
35. Ghélis C, Yon J (1982) Protein folding, Academic Press, New York, 562 p
36. Wright PE, Dyson H-J, Lerner RA (1988) Biochemistry 27:7167—7175

Received May 21, 1991
accepted August 9, 1991

Authors' address:

Prof. Dr. Rainer Jaenicke
Institut für Biophysik und Physikalische Biochemie
Universität Regensburg
Universitätsstr. 31
D-8400 Regensburg, FRG

Progress in Colloid & Polymer Science

Progr Colloid Polym Sci 86:30—35 (1991)

Analysis of intra- and intermolecular interactions in some oxidases using an analytical ultracentrifuge

J. Behlke, A. Knespel, R. W. Glaser[1]), and K.-P. Pleißner

Central Institute of Molecular Biology, Department of Hydrodynamics, Berlin, and[1]) Humboldt-University, Division of Medicine, Berlin, FRG

Abstract: The molecular mass of glutamate oxidase from Streptomyces endus was determined by means of sedimentation equilibrium experiments, sedimentation velocity and diffusion studies using an analytical ultracentrifuge. The values obtained vary between 73000 and 98000 in the concentration range of 0.05—0.65 mg/ml. By fitting the concentration-dependent point molecular masses to different models, the best approximation to the experimental data was obtained assuming a monomer-dimer equilibrium of the enzyme. The monomer molecular mass was estimated to be 57000 ± 1800 and the equilibrium constant amounts, $K_d = 3.0 \cdot 10^{-6}$ M. — Glucose oxidase from Penicillium notatum and horseradish peroxidase in an equimolar mixture at pH 5.5 are able to form a 1:1 complex with an association constant $K_a = 1.0 \cdot 10^5$ M^{-1}. This behavior can be deduced from sedimentation equilibrium runs. In the presence of $(NH_4)_2SO_4$ or other lyotropic salts the values of the association constant can increase slightly. This fact should be important for preparing bi- or multienzymes of soluble proteins.

Key words: Analytical ultracentrifugation; enzyme dissociation; equilibrium constants; bienzyme formation; lyotropic salts

Introduction

Proteins can be considered as fluctuating particles in thermal equilibrium with their environmental solvent. The motion of atoms or groups of atoms results in changes of their overall dimensions and produced channels or cavities which may be occupied by solvent or small solute molecules [1]. In general, the solvent environment leads to a reduction of motional amplitudes of groups in the protein as a result of molecular dynamics calculations [2, 3]. Changes in the hydration of proteins or their solvent structure can modify the intra- or intermolecular forces in proteins or between them [4, 5]. These effects are observable in concentration-dependent molecular mass determinations (dissociation-association phenomena) of oligomeric enzymes and partially from differences in their biological activity.

The analytical ultracentrifuge is still the method of choice to study quantitatively the macromolecular interactions under different solution conditions. In this study we communicate some methodical experiences obtained in experiments that analyzed the dissociation process of glutamate oxidase of Streptomyces endus or in the heterologous association of different proteins, i.e., horseradish peroxidase and glucose oxidase of Penicillium notatum.

Material and methods

Glutamate oxidase (GLOD) from Streptomyces endus was isolated and purified as described in [6]. The protein was dissolved in 50 mM sodium phosphate, pH 7.0 and dialyzed against the same buffer. The enzyme concentration was determined using an absorption coefficient $A^{1\%}_{275nm} = 12$.

Horseradish peroxidase (POD) was purchased from Boehringer, Mannheim (FRG), and dialyzed against 50 mM sodium acetate buffer, pH 5.5. The purity was examined according to Shannon et al. [7]. Glucose oxidase

(GOD) of Penicillium notatum was purified as published in [8]. The enzyme was dialyzed against 50 mM sodium acetate buffer, pH 5.5. An absorption coefficient $A^{1\%}_{280nm} = 15.0$ was used to determine the concentration of GOD.

Molecular mass determinations were performed using an analytical ultracentrifuge Spinco E with schlieren —, interference- or UV-optics, monochromator and photoelectric scanner, respectively. The data were obtained either by sedimentation velocity experiments in combination with diffusion measurements or using sedimentation equilibrium runs.

To work economically in terms of time and substances, most of the sedimentation and diffusion experiments were carried out in one experiment using a double sector synthetic boundary cell. About 130 µl protein solution was overlayed with dialysis buffer at low speed (maximally 6000 rpm). From the time-dependent broadening of the boundary registered at the wave length of the protein absorption band, the diffusion coefficients were calculated and corrected to the viscosity of water at 20°C. Immediately after finishing the diffusion experiment, highvelocity runs were carried out to obtain sedimentation coefficients from the moving boundary. The data obtained were corrected to viscosity and density of water at 20°C.

The partial specific volume ($\bar{v}$) of the proteins was determined either from concentration-dependent density measurements using a precision densitometer DMA 10 (Anton Paar, Graz) [9] or by calculation from the amino acid composition and the density increments according to Cohn and Edsall [10].

For sedimentation equilibrium runs, the high speed or meniscus depletion technique with column height of about 3 mm has been used. A further reduction of the time to reach the sedimentation equilibrium was possible by the overspeed procedure. The concentration distribution curves were digitalized semiautomatically with a computer-assisted device for scans as published earlier [11], or by an image processing systems BVS 6471 consisting of a CCD camera type C222 in connection with a graphic control display G 83 (ZKI) and a computer SM 1420 for analyzing the interference fringes from photographic plates.

Determination of the fringe displacement or the coordinates of the data pair $Y = f(r)$ is based on the comparison of a calculated sinus function of a period of the fringe distance with the experimental sinus function in grey values from the fringes in perpendicular direction. Each column of the arbitrarily determined window with 5—10 fringes has to be brought in correlation with the comparison function. The maximum of the correlation function has to be determined by a squared regression over a range of seven points. By the two fixed points in X-axes (r_m, r_b) with known radii, all X-values can be calculated in radii data. Simultaneously, also the magnification factor in Y-direction is obtained.

From the concentration (c) distribution or fringe displacement as a function of r the molecular masses of the proteins were determined by the formula:

$$M = \frac{2RT}{(1 - \rho\bar{v})\omega^2} \cdot \frac{d\ln c}{d(r^2)} , \qquad (1)$$

with R being the gas constant, T the absolute temperature, ρ the solvent density, and ω the angular velocity.

Dissociating proteins are recognizable by their nonlinear $\ln c/r^2$ plot. With increasing values of r or c, respectively, an enlargement of $d\ln c/d(r^2)$ and, therefore, molecular masses is obtained. The point molecular masses are weight average data:

$$M_w = \Sigma c_i \cdot M_i / \Sigma c_i , \qquad (2)$$

which vary between the data of the dissociated and associated proteins. According to Eq. (2), they can be considered as a mixture of different species with the molecular mass M_i and their partial concentration c_i. A dissociating protein consisting of two equal monomers can be expressed by the equation:

$$M_w = M_m \left(c_m + \frac{2c_m}{K} \right) / c_0 , \qquad (3)$$

where M_m and c_m denote the molecular mass and concentration of monomers, c_0 is the total concentration, and K the equilibrium constant between monomeric and dimeric protein molecules. It can be calculated directly from the concentration dependence of M_w-values only when M_m is known. M_m can be estimated by minimizing the function

$$F = \Sigma w_i [M_w^{theo} \cdot (c_0^{(i)}) - M_w^{exp}(c_0^{(i)})]^2 , \qquad (4)$$

with the procedure VA 13A of the Harvell Subroutine Library. M_w in formula (4) was calculated by Eqs. (3) and (5):

$$c_m + \frac{c_m^2}{K} - c_0 = 0 . \qquad (5)$$

By means of low weights w_i in formula (4) the influence of uncertain data i can be reduced.

To demonstrate the interaction or partial complex formation of two different proteins the following,

$$iA + jB \rightleftharpoons A_iB_j , \qquad (6)$$

and the use of interference optics for recording the concentration distribution at sedimentation equilibrium is advantageous. For this purpose it is suitable to apply a six-channel cell (Fig. 1) and to put the pure components A and B in the middle (m) and outer (o) compartment, whereas the inner (i) one contains an equimolar mixture of both proteins. From the radius-dependent fringe displacement or its transformed concentration values $c = f(r)$, obtained at the point r_0 where initial and equilibrium concentrations are equal (Eq. (7)):

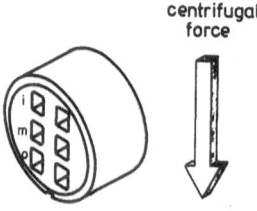

Fig. 1. Schematic representation of a six-channel cell where (*i*), (*m*) and (*o*) are the inner, middle, and outer compartment, respectively, with regard to the centre of revolution (opposite direction to the centrifugal force)

$$c_r = \Sigma c_{0,i} \cdot e^{F i M i (r^2 - r_0^2)}, \qquad \text{with} \qquad (7)$$

$$F = \frac{(1 - \rho \bar{v}) \omega^2}{2RT}, \qquad (8)$$

the molar mass M of component i was calculated by fitting of the experimental data pairs $c = f(r)$ to the theoretical ones.

In case of a protein mixture the concentration distribution was analyzed as a sum of two or more exponential functions. Considering the molecular masses of the single proteins or the complex(es), their partial concentrations c_i at distinct points of r were determined by fitting of the experimental data $c = f(r)$ according to Eq. (7). From the partial concentration the association constants were calculated depending on the conditions used in the experiments.

Results

1) Glutamate oxidase

The enzyme is characterized by sedimentation coefficients of 5—6S with lower values for the more diluted protein samples (Fig. 2). The diffusion coefficients of glutamate oxidase amount to about $6 \cdot 10^{-7}$ cm²/s. These data increase slightly when the protein is diluted continuously (Fig. 2). Using a partial specific volume $\bar{v} = 0.735$ cm³/g (see Table 1), from sedimentation and diffusion coefficients a molar mass of 80000—95000 g/mol was calculated. The lower data obtained in more diluted samples suggest a partial dissociation of the enzyme.

To examine these results by an independent method, sedimentation equilibrium runs were carried out. A non-linear $\ln c/r^2$ plot (not shown) indicates the existence of an association-dissociation equilibrium. Point molecular weights result in data

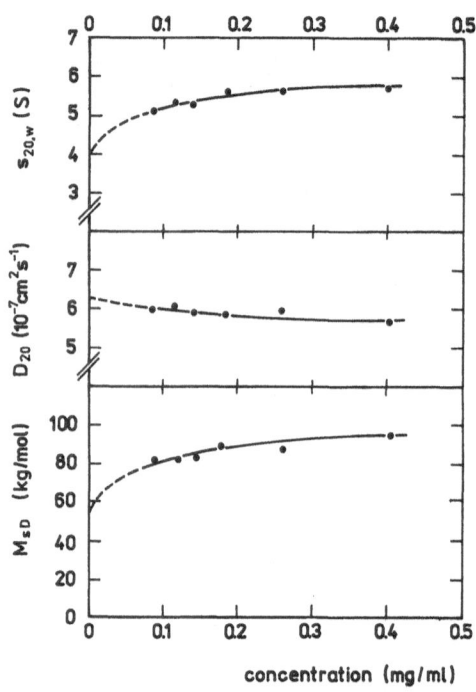

Fig. 2. Sedimentation and diffusion coefficients as well as molecular masses of glutamate oxidase in dependence on the protein concentration. Buffer: 50 mM sodium phosphate, pH 7.0

Table 1. Partial specific volume of the oxidases obtained as mentioned in "material and methods"

Enzyme	$\bar{v}$ (cm³/g)
Glutamate oxidase	0.735[1]
Peroxidase	0.718[2]
Glucose oxidase	0.725[1]

[1]) calculated; [2]) experimentally determined

of 62000—71000 at 0.04 mg/ml and about 98000 at 0.65 mg/ml (Fig. 3). These concentration-dependent weight average molecular weights were fitted to test different models. The best approximation of the experimental data was obtained when assuming a monomer-dimer equilibrium using Eqs. (3)—(5). Because neither total dissociation of the dimeric molecules into monomeric ones, nor their full

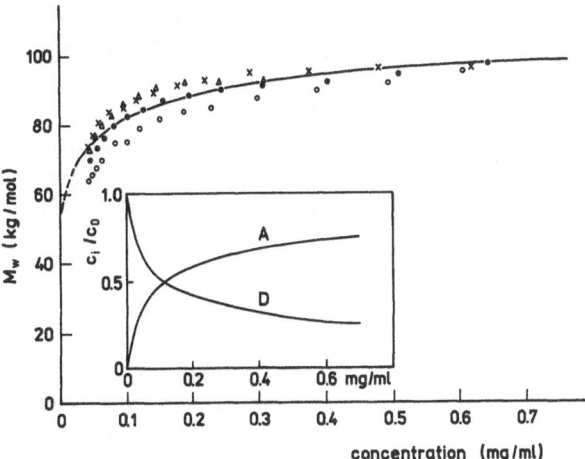

Fig. 3. Point molecular weights of glutamate oxidase in dependence on the concentration derived from a sedimentation equilibrium run usig an AnG rotor. The symbols designate the different initial concentrations in the cells. The insert demonstrates the partial concentration of the associated (*A*) and dissociated (*D*) protein molecules assuming a monomer dimer equilibrium and an equilibrium constant $K_d = 3.0 \cdot 10^{-6}$ M

association was observed under these experimental conditions, the value M_m was estimated from the data pairs $M_w = f(c)$ as described in "materials and methods". Considering the data of four different experiments (Fig. 3), M_m amounts to 57200 ± 1800. Taking this value and assuming two equal monomeric subunits in the glutamate oxidase, an average equilibrium constant $K_d = 3.0 \cdot 10^{-6}$ M was calculated. This value allows calculation of the content of monomeric and dimeric glutamate oxidase in dependence on the total concentration (see insert in Fig. 3). Considering the concentration-dependent molecular masses obtained from the sedimentation and diffusion coefficients (Fig. 2) an average equilibrium constant $K_d = 2.7 \cdot 10^{-6}$ M was determined.

2) Peroxidase — glucose oxidase interaction

The molecular mass of peroxidase and glucose oxidase dissolved in 50 mM acetate buffer pH 5.5 were determined by the sedimentation equilibrium technique using a six-channel cell (Fig. 1). By fitting the experimental determined values of the radius-

dependent concentration distribution for a single protein using Eq. (7), molecular masses of 37000 ± 3000 and 142000 ± 3000 were determined for the peroxidase and glucose oxidase, respectively. Both proteins were homogeneous and showed no tendency to form aggregates or to contain degradation products. When analyzing the concentration distribution of an equimolar mixture of both proteins obtained under the same sedimentation equilibrium conditions, there was no optimal fit when taking into account only the single proteins with their molecular masses M_i, as mentioned before. The adaptation to the experimental curve was fairly good under consideration of an additional compound of higher molecular weight. A more optimal fit was achieved by a third component of about 180000 without excluding any other higher molecular species. Thus, we can assume that a distinct part of the peroxidase and glucose oxidase molecules is able to form a complex, presumably in a 1:1 stoichiometry. In Fig. 4 an experimental curve containing about 16% peroxidase — glucose oxidase complex, as well as an expected curve without complexing are presented. The tendency to complex formation between the two enzymes is obviously not strong. Because the isoelectric points differ considerably (glucose oxidase: IP ~ 4.5, peroxidase IP ~ 7.2) at pH 5.5, one protein exists as a polyanion (glucose oxidase) and the other as a polycation (peroxidase). Electrostatic forces could be responsible for their complex formation.

Further studies on the interaction between the two proteins were carried out in the presence of lyotropic electrolytes, especially $(NH_4)_2SO_4$. As demonstrated in Fig. 5, an increase of the concentration of $(NH_4)_2SO_4$ up to 0.1 M promotes the complex formation, whereas higher amounts of salt lead to a destabilization of the heterologous interaction. Similar, but not so distinct effects were observed also with Na_2SO_4, NaCl, and KCl.

Discusssion

The results obtained with the analytical ultracentrifuge yield new insights into the structure and solution behavior of glutamate oxidase. Instead of the apparently different molecular mass deduced from gel filtration experiments with 90 kDa or polyacrylamide gel electrophoresis with 50 kDa [6], we could demonstrate the enzyme to be a dimeric protein which is able to dissociate during con-

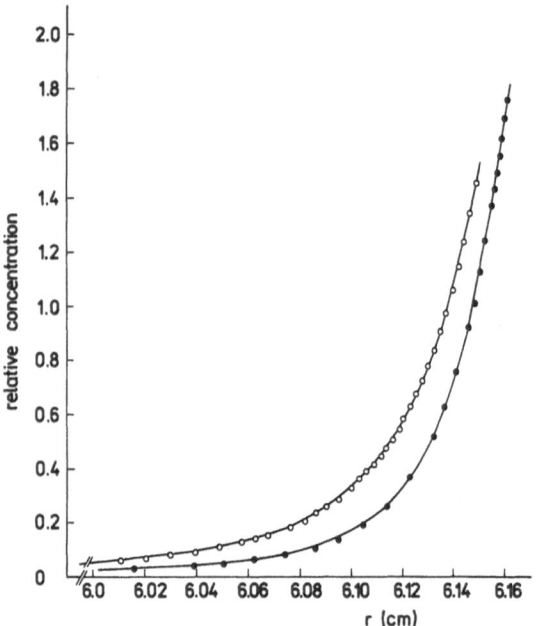

Fig. 4. Concentration distribution $c = f(r)$ of an equimolar mixture of $2.16 \cdot 10^{-6}$ M glucose oxidase and peroxidase dissolved in 50 mM sodium acetate buffer, pH 5.5 at sedimentation equilibrium; speed 20000 rpm, temperature 12.5 °C. (●) Experimental data corresponding to 16% complexed enzyme ($K_a = 1.1 \cdot 10^5$ M^{-1}), (○) calculated curve of a protein mixture without complex formation

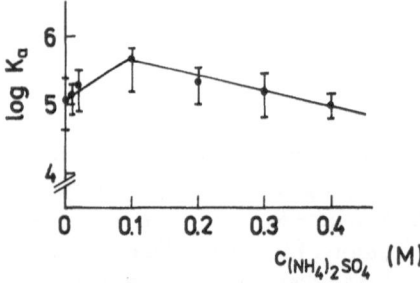

Fig. 5. Influence of $(NH_4)_2SO_4$ on the complex formation between glucose oxidase and peroxidase. Conditions as given in Fig. 3

tinuous dilution with buffer. This was possible by means of the sedimentation equilibrium technique and analysis of the concentration dependent point molecular masses, as well as by combined diffusion and sedimentation velocity experiments using a synthetic boundary cell. The concentration depend-

ence of the molecular masses obtained by the two independent methods and, thus, the equilibrium constants are in good accordance.

Using the equilibrium constant it is possible to determine the partial concentration of the different associated or dissociated components in oligomeric proteins depending on the initial concentration. These results can be compared with the kinetic data of the substrate metabolism of the enzymes to recognize whether there are differences in the activity of different associated species in the oligomeric proteins or not. Because no such differences were observed in glutamate oxidase (Dr. H. Honeck, personal communication) both subunits are equivalent and work independently of one another. This is not the case with the D-amino acid oxidase [12] for which a monomer-dimer equilibrium and higher activities for the monomers have been described. Although neither the pure monomeric nor the dimeric enzyme must be present, its kinetic data can be estimated from at least two data pairs of concentration dependent activity values and the equilibrium constant. Recently, we were able to recognize the active tetrameric phosphofructokinase which was in equilibrium with the inactive monomeric and dimeric species from two equilibrium constants and at least three different data pairs of the concentration-dependent activity values [5].

Using this enzyme we could also demonstrate the promoting effect of the lyotropic anions, sulfate and phosphate, on the reassociation of the partially dissociated enzyme with restoration of the activity.

Obviously, this behavior of $(NH_4)_2SO_4$ is also of importance for the heterologous association of peroxidase and glucose oxidase (bienzyme formation). Both proteins were used for the quantitative determination of glucose by oxidation to gluconate in the presence of glucose oxidase. The additionally formed H_2O_2 is the substrate in the peroxidase catalyzed signal reaction to transform a chromogene in a dye. According to [13], the velocity of a combined substrate metabolism is higher when different enzymes form complexes. Such a behavior could be demonstrated in equimolar mixtures of both proteins when adding $(NH_4)_2SO_4$. Under the conditions used in the experiments, 0.1 M $(NH_4)_2SO_4$ had an optimal effect with regard to complex formation (see Fig. 5) and the speed of glucose oxidation (Dr. F. Dittrich, personal communication).

About the reason for such an association promoting effect, one can only speculate. According to

[4], lyotropic anions such as sulfate or phosphate anions generally favor the formation of a water shell around the protein. This affects a conformational change to a more globular shape, which is obvioulsy of advantage for complexing. Increasing concentrations of this lyotropic salt should have an opposite effect, because both proteins exist either as a polyanion (GOD) or a polycation (POD), and the possible electrostatic contacts between them should be weakened. If the formation of bienzymes in an equilibrium reaction takes place, the amount of heterologous complexes can be increased by using a higher protein concentration. The other but more economical possibility is to optimize the addition of lyotropic salts to increase the association constant.

Acknowledgements

We are indebted to Drs. H. Honeck and K. Welfle for purification of glutamate oxidase and glucose oxidase, respectively.

References

1. Cooper A (1976) Proc Natl Acad Sci (USA) 73:2740—2741
2. van Gunsteren WF, Karplus M (1982) Biochemistry 21:2259—2274
3. van Gunsteren WF, Berendsen HJC, Hermans J, Hol WGJ, Postma JPM (1983) Proc Natl Acad Sci (USA) 80:4315—4319
4. Arakawa T, Timasheff SN (1984) Biochemistry 23:5912—5923
5. Bär J, Huse K, Kopperschläger K, Behlke J, Schulz W (1988) Int J Biol Macromol 10:99-105
6. Böhmer A, Müller A, Passarge M, Liebs P, Honeck H, Müller HG (1989) Eur J Biochem 182:327—332
7. Shannon LM, Kay E, Lew JY (1966) J Biol Chemistry 241:2166—2172
8. Welfle K, Büttner W, Behlke J (1990) Studia Biophys 138:245—260
9. Behlke J (1971) Studia Biophys 28:79—84
10. Cohn EJ, Edsall JT (1943) Proteins, amino acids and peptides, Academic Press, New York
11. Pleißner K-P, Wessel R, Knespel A, Meissner F, Behlke J (1986) Exper Technik Physik 34:139—145
12. Yagi K, Sigiura N, Ohama H, Ohishi N (1973) J Biochem 73:709—714
13. Gaertner FH (1978) Trends in Biochem Soc 5:63—65

Authors' address:

Prof. Dr. J. Behlke
Zentralinstitut für Molekularbiologie
Robert-Rössle-Str. 10
O-1115 Berlin, FRG

Progress in Colloid & Polymer Science   Progr Colloid Polym Sci 86:36—40 (1991)

Sedimentation velocity analysis of the DNA-protamine interaction

S. Bickhardt, G. Ebert[1]), N. Nishi[2]), and K. Hagiwara[3])

[1]) Fachbereich Physikalische Chemie der Philipps-Universität Marburg
[2]) Department of Polymer Science, Faculty of Science, Hokkaido University, Sapporo, Japan
[3]) Department of Molecular Biology, National Institute of Agrobiological Resources,
   Tsukuba Science City, Ibaragi, Japan

Abstract: The aim of this work was to investigate the molecular weight of DNA-protamine aggregates in aqueous solution as a function of the [Arg]/[nucleotide] ratio and as a function of the molecular weight distribution of the DNA. In the case of the polydisperse DNA the ultracentrifuge and the UV-absorption measurement show an increasing molecular weight with increasing clupeine content from 2×10^6 of the DNA itself to 18×10^6 of the aggregates. On the other hand, the homodisperse DNA shows the parallel behavior, but the molecular weight of the aggregates is four times higher. The high molecular weight aggregates of both samples have a very broad molecular weight distribution according to the broad sedimentation boundary traces. These high molecular weight aggregates are supposed to form networks of DNA-protamine complexes.

Key words: Ultracentrifuge; sedimentation velocity run; DNA; clupeine; protamine; aggregation; molecular weight distribution

Introduction

Protamines are strong basic nucleoproteins with an arginine content of $\approx 70\%$ [1, 2]. The molecular weight of protamines amounts to 4000—8500 in general. All protamines are heterogeneous and consist of definite amounts of several components. Clupeine was the protamine used for this study. The three components of clupeine are illustrated in Fig. 1.

The protamines can be found mainly in mature sperm nuclei of fish and also in those of mammals; there they are supposed to preserve and protect the genetic information. However, their biological function is not yet completely understood; it is known that they replace the histones in the DNA-histone complexes during spermatogenesis [1]. The resulting DNA-protamine complex is a very condensed one, where all DNA molecules are oriented parallel to each other. In this complex the transcription process is completely inhibited, and the DNA is protected from enzymatic hydrolysis.

The formation of the DNA-protamine complex is mainly due to electrostatic interaction between the arginine residues in protamine and the phosphate residues in DNA.

In this paper molecular weights of DNA-clupeine (protamine from herring testes) aggregates in aqueous solution are determined as a function of the [Arg]/[nucleotide] ratio, and as a function of the molecular weight distribution of the DNA, by sedimentation-velocity measurements with an analytical ultracentrifuge.

The results of these experiments are compared with those of former spectrophotometric measurements for the DNA-protamine aggregate in solution [3, 4].

Experimental

Material

Clupeine sulfate (Sigma, Deisenhofen, FRG) was converted into hydrochloride with Amberlite IRA 400 (Cl

AMINO-ACID SEQUENCES OF THE THREE COMPONENTS OF CLUPEINE

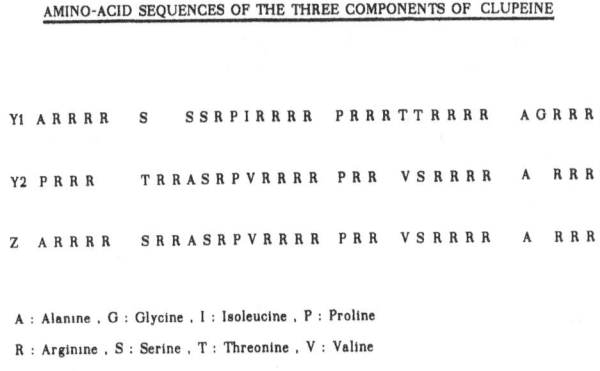

```
Y1  ARRRR   S    SSRPIRRRR  PRRRTTRRRR   AGRRRR

Y2  PRRR    TRRASRPVRRRR  PRR  VSRRRR   A RRRR

Z   ARRRR   SRRASRPVRRRR  PRR  VSRRRR   A RRRR
```

A : Alanine , G : Glycine , I : Isoleucine , P : Proline

R : Arginine , S : Serine , T : Threonine , V : Valine

Fig. 1. Amino-acid sequences of the three components Y1, Y2, and Z of clupeine, according to [1]

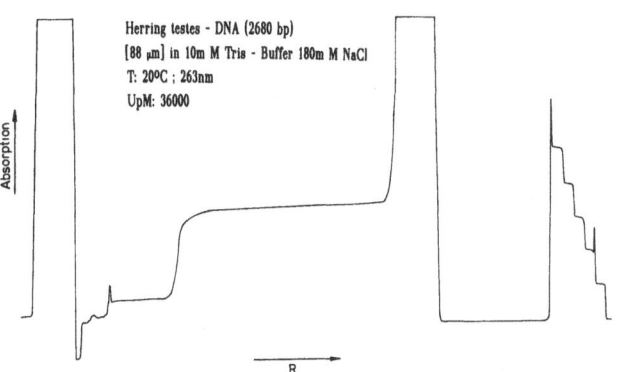

Fig. 2. Sedimentation boundary trace of herring testes DNA (2680 base pairs) (88 μm) in 10 mM Tris-Buffer pH: 7.65, 180 mM NaCl, T: 20°C, λ = 263 nm, rpm: 36000

form). The herring testes DNA Type XIV was obtained from Sigma. The homodisperse linear double-stranded DNA with 2680 base pairs was prepared by proliferation of plasmid PUC 12 in E. coli (JM 103), followed by its isolation, purification and digestion with restriction enzyme (Eco RI). High purity of the resulting DNA was confirmed by agarose-gel electrophoresis.

Tris(hydroxymethyl-)-aminomethane p.a. from Merck AG (Darmstadt, FRG) was used without further purification.

Preparation

A solution of the protamine in 10 mM Tris-buffer pH 7.65 and 180 mM NaCl was added to a DNA solution in 10 mM Tris-buffer and 180 mM NaCl by an Eppendorf pipette. Both solutions were mixed at 20°C. The concentration of Arg residues in the protamine solution was 964 μM. Concentration of nucleotide in DNA was 95.08 μM and that in the homodisperse DNA was 88.37 μM.

After gentle shaking, the solution was smoothly injected (with a 1 ml syringe) directly into a 12-mm double-sector Al-cell for a sedimentation-velocity run.

Ultracentrifugation

All measurements were done using an analytical ultracentrifuge model E (Beckman Instruments, Palo Alto, California, USA). a 12-mm double-sector-Al-cell was used for all experiments. The sedimentation-velocity runs were recorded with a UV-scanner at 263 nm and 20°C. The partial specific volume was obtained by density measurment and compared with literature values [5], the value for the partial specific volume V_2^* used was 0.55 ml/g.

Figures 2 and 3 show an example for a sedimentation boundary trace of the homo- and the polydisperse DNA; they clearly show the influence of the molecular weight distribution.

Fig. 3. Sedimentation boundary trace of herring testes DNA (polydisperse) (95 μm) in 10 mM Tris-Buffer pH: 7.65, 180 mM NaCl, T: 20°C, λ 263 nm, rpm: 36000

Results and discussion

Nishi et al. investigated the relationship between the characteristic primary structure of protamine and the function of DNA-protamine binding with absorption and CD-spectra [3, 4, 6]. Protamine as well as different model polypeptides were used for these experiments.

It was shown that the absorption of DNA-protamine complexes changes in relation to the degree of aggregate formation (Fig. 4). The absorption around 260 nm undergoes a remarkable red shift and the absorption near to 300—400 nm increases simultaneously with increasing protamine content. The average molecular weight of the aggregate can

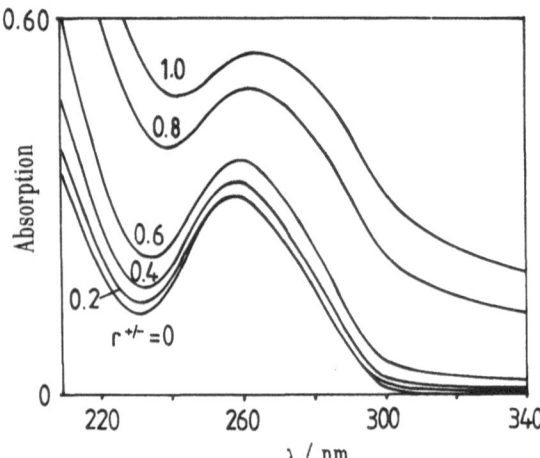

Fig. 4. Change in the absorption spectrum of DNA induced by clupeine. $r+/-$: [Arg]/[nucleotide], [nucleotide] = 50 μM [NaCl]: 100 mM, 10 mM Tris-HCl buffer, pH 7.5

be roughly estimated by the method of Clevan and Schuhmaker [3, 7] from the absorption between 320—340 nm with the following equation:

$$M = \frac{3\tau}{16\pi Kc} \quad (1)$$

$$K = \frac{2\pi^2\eta_0^2(\delta n/\delta c)^2}{N_A\lambda^4} \quad (2)$$

where M is molecular weight, c is concentration of the complex, τ is O.D. between 320—340 nm, λ is wave-length, and N_A is $6 \cdot 10^{23}$ Avogadro's Number.

In addition to the absorption measurements also studies of CD-spectra of the DNA-protamine complexes were carried out. The CD-spectra show that protamine molecules induce a DNA-aggregation very efficiently, but they do not change the DNA-conformation remarkably. The assumed conformation may be a rigid turn structure. The high efficiency in aggregation of protamine is due to the geometrical adaptibility of the double-stranded DNA.

The sedimentation velocity run in the ultracentrifuge is a further method to characterize the DNA-protamine aggregation. It is possible to calculate the molecular weight with the Flory-Mandelkern equation (below) [8] by means of the sedimentation coef-

ficient. But there is no possibility to determine the molecular weight in the range of $\approx 10^6$ directly by sedimentation equilibrium runs because one cannot apply such a low stable r.p.m. (round per minute) in the ultracentrifuge, which is necessary for such high molecular weights.

Flory-Mandelkern equation:

$$M = \left[\frac{s_{20,w}^0[\eta]^{0.5}\eta_0 N_A}{\beta(1 - {}^*V_2\rho_0)}\right]^{3/2}, \quad (3)$$

where M is molecular weight, $s_{20,w}^0$ is sedimentation coefficient, η is viscosity, N_A is Avogadro's Number, ρ_0 in density of water, and β is $2.4 \cdot 10^6$.

The sedimentation velocity runs in the ultracentrifuge were done under the same conditions as the measurements in [3, 4]. These experiments allow to compare the interaction of homo- and polydisperse DNA with clupeine. Tables 1 and 2 show the results together with the experimental conditions.

Table 1 illustrate the interaction of polydisperse DNA with clupeine. With increasing clupeine content an increasing molecular weight of the aggregate can be detected. At the highest $r+/-$ (ratio of [Arg])/[nucleotide]) value of 0.45 a molecular weight of 18×10^6 was obtained. The DNA-protamine complexes show a very broad molecular weight distribution. In comparison, Table 2 shows the results of the interaction of the homodisperse DNA with clupeine. There are obviously some important differences: one can just reach a $r+/-$ value of 0.4, and the molecular weight at this $r+/-$ value is more than four times higher that of the polydisperse DNA/protamine complexes. But one can also notice a wide molecular weight distribution of the high-molecular-weight DNA-protamine aggregates.

The molecular weights of the two DNA samples lie approximately in the same region, close to 2×10^6. This fact is very important for comparison of the two substances. The clupeine ratio exerts a different influence on the molecular weight of the complex, depending on the heterogeneity of the DNA. If the homodisperse DNA is used for the experiments, larger aggregates with a higher molecular weight are found. These results clearly show the influence of the dispersity of the DNA.

Remarkably, in the case of the homodisperse DNA only 30—40% of the sedimenting molecules consists of highmolecular weight DNA-protamine aggregates, whereas the remaining fraction seems

Table 1. Determination of the interaction of polydisperse DNA and Clupeine HCl

Measuring conditions:

T:	20 °C
λ:	263 nm
cell:	Aluminum-double sector
DNA-concentration:	95.08 µM in 10 mM Tris-buffer and
(before mixing)	180 mM NaCl
Protamine-concentration:	964 µM in 10 mM Tris-buffer and
(before mixing)	180 mM NaCl

$r+/-$	RpM [min^{-1}]	$(S_{20,w})_c$ [s]	M
0	36000	$10.02 \cdot 10^{-13}$	1800000 [±200000]
0.2	24000	$13.00 \cdot 10^{-13}$	2400000 [±200000]
0.3	14000	$27.07 \cdot 10^{-13}$	4900000 [±500000]
0.35	10000	$54.40 \cdot 10^{-13}$	10000000 [±1000000]
0.4	6400	$89.80 \cdot 10^{-13}$	16000000 [±1500000]
0.45	6400	$98.44 \cdot 10^{-13}$	18000000 [±1500000]

Table 2. Determination of the interaction of DNA (2680 base pair) and Clupeine HCl

Measuring conditions:

T:	20 °C
λ:	263 nm
cell:	Aluminum-double sector
DNA-concentration:	88.37 µM in 10 mmol Tris-buffer and
(before mixing)	180 mM NaCl
Protamine-concentration:	964 µM in 10 mmol Tris-buffer and
(before mixing)	180 mM NaCl

$r+/-$	RpM [min^{-1}]	$(S_{20,w})_c$ [s]	M
0	36000	$13.27 \cdot 10^{-13}$	2400000 [±200000]
0.1	28000	$11.72 \cdot 10^{-13}$	2100000 [±200000]
0.2	28000	$13.10 \cdot 10^{-13}$	2400000 [±200000]
0.25	8000*)	$148.00 \cdot 10^{-13}$	27000000 [±3000000]
0.3	14000	$14.45 \cdot 10^{-13}$	2600000 [±200000]
	3600*)	$156 \cdot 10^{-13}$	28000000 [±3000000]
0.35	20000	$12.60 \cdot 10^{-13}$	2300000 [±2000000]
	3000*)	$295 \cdot 10^{-13}$	54000000 [±5000000]
0.4	3000*)	$487 \cdot 10^{-13}$	88000000 [±9000000]

*) High molecular weight fraction ≈ 30—40%.

to be DNA itself. The measurement of a higher $r+/-$ ratio than 0.45, respectively 0.4, was impossible because turbidity occurs during the acceleration in the ultracentrifuge. (Turbidity also occurs by rapid stirring or shaking of the solution.) Therefore, it is difficult to answer the question of whether a higher protamine ratio will lead to a higher percentage of high-molecular-weight aggregates. Further experiments in the future, for example using electron microscopy, will further elucidate the aggregates.

Figure 5 shows a comparison of the experimental results according to [3] and to the ultracentrifuge runs. One can see that the general tendency of the DNA-protamine aggregation, which was at first shown by UV measurements, can be confirmed with the help of ultracentrifuge experiments. The interaction of the polydisperse DNA with protamine resembles Nishi's results which were also obtained by experiments with the same polydisperse DNA. In contrast, however, the homodisperse DNA shows a different behavior.

The molecular weight of the used DNA samples lies in the region of 2×10^6. This is equivalent for approximately 3000 base pairs and 6000 phosphate residues. The arginine residues of clupeine interact with the phosphate residues of the DNA. If the arginine residue of the protamine molecule is bound by one phosphate residue of the DNA the calculated molecular weight of the "primary complex" is 2.5×10^6. The clupeine molecules are supposed to be fixed in the small groove of the DNA. This idea is supported by fiber diffraction studies [9].

This complex is formed at a low [Arg]/[nucleotide] ratio (0.1—0.2). If the clupeine content increases, the molecular weight of the aggregates

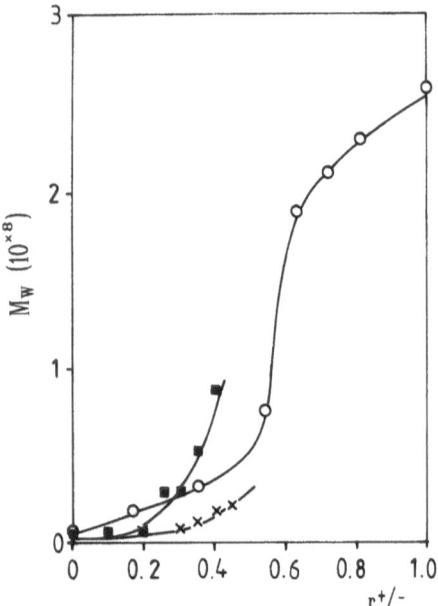

Fig. 5. Molecular weight of DNA aggregate induced by clupeine $r+/-$: [Arg]/[nucleotide]. [NaCl]: 180 mM, 10 mM Tris-HCl buffer, pH 7.65; ○: Measurement of Nishi et al.[3]), [Nucleotide]: 50 μM; ×: polydisperse DNA, [Nucleotide]: 95 μM; ■: DNA (2680 basepair), [Nucleotide]: 88 μM

In the "primary complex" the clupeine molecules wind around the small groove of the DNA. One can support the thesis that the ends of the clupeine chains may overlap on the DNA double helix. These overlaping ends might be able to join together some DNA helices, and this can be an explanation for the networks formed that have a very high molecular weight.

Future experiments should help to clarify the function of each clupeine component. Electron microscopy studies are being carried out at present; they may give a hint to the structure of the complexes.

shows the same tendency. This leads to the model of the "secondary complex" where several primary complexes are joined together in places by protamine. At a higher [Arg]/[nucleotide] ratio close to 1 the UV-measurements of Nishi show a molecular weight of 250×10^6; these results indicate the building of stable networks of "secondary complexes".

By ultracentrifuge runs it is not possible to measure $r+/-$ values higher than 0.45 because of turbidity. But the complex of the polydisperse DNA with protamine at a $r+/-$ value of 0.45 has a molecular weight of 18×10^6; this result is completely comparable to Nishi's ones, whereas the molecular weight of the aggregate of the homodisperse DNA with protamine at a $r+/-$ value of 0.40 is four times higher than that of the polydisperse one. This is also a clue to the formation of a network if one uses homodisperse DNA.

References

1. Ando T, Yamasaki M, Suzuki K (1973) Protamines-Characterization, Structure and Function. Springer-Verlag, Berlin
2. Willmitzer L, Wagner KG (1980) Biophys Struct Mech 6:95—110
3. Nishi N, Ogawa K, Hayasaka H, Hasegawa K (1990) Peptide Chem, pp 325—330
4. Nishi N (1988) Peptide Chem 1987, 747—750
5. Aten JBT, Cohen JA (1965) J Mol Biol 12:537—548
6. Nishi N, Tsunemi M, Hagiwara K, Tokura S, Tsutsumi A (1986) Peptide Chem 1985, 245—250
7. Clevan L, Schumaker VN (1982) Nucl Acids Res 10:6809—6817
8. Eigner J (1968) "Methods in Enzymology Vol XII", Academic Press, New York, London, 386—429
9. Saenger W (1988) "Principles of Nucleic Acid Structure", Springer Verlag, New York, 400—402

Author's address:

Prof. Dr. G. Ebert
Fachbereich Physikalische Chemie — Biopolymere —
Philipps Universität Marburg
Hans-Meerwein Str., Gebäude H
W-3550 Marburg (Lahn), FRG

Comparative determination of the particle weight of glycoproteins by SDS-PAGE and analytical ultracentrifugation*)

H. Durchschlag, P. Christl, and R. Jaenicke

Institute of Biophysics and Physical Biochemistry, University of Regensburg, Regensburg, FRG

Abstract: The particle weight of 21 glycoproteins has been determined by SDS-PAGE and high-speed sedimentation equilibrium, focussing on the influence of carbohydrate content, partial specific volume ($\bar{v}_2$), and relative molar mass (M_r) on the results of both techniques. Results were compared with literature data and with 22 nonconjugated proteins yielding the following results: (i) $\bar{v}_2$ of native glycoproteins decreases linearly with increasing carbohydrate content. Therefore, $\bar{v}_2$-values of glycoproteins not accessible to experimental determination may be calculated. (ii) The determination of M_r of glycoprotein subunits by SDS-PAGE shows anomalies in the electrophoretic behaviour. Small amounts of impurities have no influence on the result, if one refers only to the major band(s). Systematic variation of the electrophoretic parameters T and C, as well as use of Ferguson plots only leads to insignificant improvements. There is no correlation between the carbohydrate content and anomalies in the electrophoretic behaviour. A possible explanation for this behaviour may be found in different conformations of the sugar moiety, different interactions with the electrophoresis gel, and different extent of SDS binding. (iii) Analytical ultracentrifugation yields accurate values for M_r of native glycoproteins, independent of the carbohydrate content and the size of the molecule. The application of this technique, however, requires the exact knowledge of $\bar{v}_2$ and the absence of impurities. (iv) A comparison of the molar subunit masses obtained by the two methods with literature values clearly proves the superiority of analytical ultracentrifugation, when compared with SDS-PAGE.

Key words: Glycoproteins; molar mass; comparative studies; SDS-PAGE; partial specific volume; analytical ultracentrifugation

1. Introduction

1.1. Simple and conjugated proteins

Based on their composition, proteins can be classified as "simple" or "conjugated". Upon hydrolysis, simple (nonconjugated) proteins yield only amino acids, whereas conjugated proteins in addition contain inorganic or organic components as prosthetic groups. On the basis of the chemical nature of these groups, conjugated proteins may be subdivided into nucleo-, lipo-, glyco-, phospho-, hemo-, flavo-, and metalloproteins. The percentage by weight of nonprotein components in different conjugated proteins varies in a wide range, from less than 1% to as high as 90% in the case of some glyco- and lipoproteins.

1.2. Glycoproteins

Within the family of conjugated proteins, the glycoproteins represent a major group of wide distribution and high biological significance.

---

*) Dedicated to Prof. Dr. Dr. h.c. mult. Otto Kratky on the occasion of his 90th birthday.

Among the glycoproteins, enzymes, hormones, and the immunoglobulins are found; glycoproteins may occur in the blood plasma, urine, egg white, in addition they are constituents of mucus, connective tissue and membranes. The different types of glycoproteins have a number of important biological tasks (e.g., structural support, enzymatic catalysis, transport, signal transduction, nerve transmission, recognition, memory, blood clotting, agglutination, cell adhesion, lubrication, (immuno)protection, defence, control, antifreeze activity).

As in the case of simple proteins, the molar masses of glycoproteins vary between $\simeq 10$ and $> 10^3$ kg/mol. Their carbohydrate content ranges from $< 1\%$ to $> 80\%$, with maximum values in certain mucoproteins and blood group substances.

Glycoproteins contain carbohydrate groups covalently attached to the polypeptide chain. Hexoses (e.g., mannose, galactose) and hexose derivatives (e.g., N-acetylhexosamines) are the major constituents of the carbohydrate moiety. The side chains of glycoproteins may be linear or branched, containing a few to dozens of monosaccharide residues. The terminal unit often carries a negatively charged residue (sialic acid) or L-fucose.

The exact topography of the carbohydrate residues with regard to the protein moiety of glycoproteins is not fully resolved yet. In general, the protein part is inside (protein core), and the carbohydrates outside. The carbohydrate residues may be arranged in form of antennae on the protein, thus representing a flexible and expanded structure whose precise conformation may be influenced by the ionic strength of the environment. In other cases, the glycan chains may cover parts of the protein surface.

There are examples where the sequence of the carbohydrate moiety has been shown to vary, due to post-translational processing ("tailoring"). This microheterogeneity proves clearly that glycosylation is not controlled genetically. The biosynthesis of the carbohydrate chains is catalyzed by enzymes (glycosyl transferases), whereas a series of other enzymes (glycohydrolases) is able to disrupt the carbohydrate residues. When analysing glycoproteins, this property may be used for cleaving sugar moieties.

For detailed reviews on the special features of glycoproteins the reader is referred to the literature (e.g., [1—4]).

1.3. Problems encountered with glycoproteins

As shown by a series of data in the literature, conflicting results for many molecular parameters of glycoproteins (like mass, carbohydrate content, absorption coefficient) have been reported. Reasons for this behaviour may comprise: lack of purity, (micro)heterogeneity with respect to the content and structural arrangement of the carbohydrate, lack of auxiliary data such as concentration (c_2) and partial specific volume ($\bar{v}_2$) as necessary prerequisites for the determination of the mass.

Efficient methods to determine the mass of proteins are polyacrylamide gel electrophoresis (PAGE) under denaturing conditions (e.g., in the presence of sodium dodecyl sulfate (SDS)), and analytical ultracentrifugation (AUC) under native or denaturing conditions (e.g., in the presence of guanidinium chloride, urea, pH). These methods yield the mass of either the native molecule or the subunit(s) (SU). While AUC supplies absolute values for the molar mass (in g/mol), M_2, of proteins, application of PAGE necessitates use of calibration proteins, thus delivering only relative molar masses (dimensionless quantities), M_r. The numerical values of both types of molar masses, however, are the same.

Nowadays, in biochemical work, SDS-PAGE (cf. [5—7]) is most commonly applied to determine the relative molar subunit mass, $M_{r,SU}$, mainly because the method is fast and simple. The procedure has been used for both nonconjugated proteins and glycoproteins, despite the irregularities encountered with conjugated proteins. Thus, for different conjugated proteins (lipo-, nucleo-, glycoproteins), as well as for proteins with high net charge (e.g., histones, as well as membrane proteins and proteins with unusual amino acid composition) manifold erroneous results for $M_{r,SU}$ have been reported in the literature. Specifically for glycoproteins, where generally anomalous (too high) $M_{r,SU}$-values are found, reference is made to [5, 7—15].

For the precise determination of M_2 from ultracentrifugal data (cf. [16—18]), exact knowledge of $\bar{v}_2$ is required. The value of $\bar{v}_2$ also influences the electrophoretic behaviour of proteins in SDS-PAGE, insofar as the extent of SDS binding to the protein, as well as hydration phenomena may play a role in the size and shape of the protein solubilized by the detergent.

1.4. Goals

The aim of the present work is (i) the determination of M_r of glycoproteins by means of SDS-PAGE and AUC, and (ii) the search for correlations between the molar mass, the partial specific volume and the carbohydrate content of glycoproteins. In this context, the exact analysis of both the protein and carbohydrate content of selected glycoproteins, and the partial specific volume, $\bar{v}_2$, is required. In addition, the detailed investigation of M_r includes the systematic variation of experimental parameters, such as gel concentration or cross-linking in the case of SDS-PAGE, and variation of speed in the case of AUC.

In order to correlate certain anomalies of glycoproteins with the carbohydrate content, experiments were paralleled by controls using simple proteins.

2. Materials

For the following experiments, selected glycoproteins and simple proteins were used (Tables 1 and 2). Enzymes suitable for the "processing" of glycoproteins, endoglycosidase F/N-glycosidase F from *Flavobacterium meningosepticum* were obtained from Boehringer (Mannheim), D-(+)-mannose and the Ninhydrin Reagent Solution from Sigma (Munich). All other reagents were of analytical grade, preferably from Merck (Darmstadt) or Serva (Heidelberg).

3. Methods

3.1. Chemical analyses and procedures

a) Protein determination: The protein content of glycoproteins was measured using either the method of Bradford [21] or Moore [22]. Calibrations made use of

Table 1. Survey of selected nonconjugated proteins, arranged alphabetically

Abbreviation	Protein	Source	Manufacturer/Reference[a]
a	Apoferritin	horse	Σ
c	Catalase	bovine	B
ch	Chymotrypsinogen A	bovine	S
cs	Citrate synthase	porcine	B
cy	Cytochrome c	horse	R
f	Ferritin	horse	B
g	Glyceraldehyde-3-phosphate dehydrogenase	yeast	[19, 20]
hb[b]	Hemoglobin	horse	S
i[c]	Insulin	bovine	Σ
la	Lactalbumin	bovine	S
l	Lactate dehydrogenase	porcine	B
βl	β-Lactoglobulin	bovine	S
ly	Lysozyme	chicken	S
mb	Myoglobin	whale	S
3p	3-Phosphoglycerate kinase	yeast	B
pa	Phosphorylase a	rabbit	B
pb	Phosphorylase b	rabbit	B
r	Ribonuclease A	bovine	Σ
b	Serum albumin	bovine	S
h	Serum albumin	human	S
s	Superoxide dismutase	bovine	B
t	Trypsin	bovine	S

[a] B: Boehringer (Mannheim); R: Roth (Karlsruhe); S: Serva (Heidelberg); Σ: Sigma (Munich).
[b] hb(α): α-chain; hb(β): β-chain.
[c] i(A): A-chain; i(B): B-chain.

Table 2. Survey of selected glycoproteins, arranged alphabetically

Abbreviation	Protein	Source	Manufacturer[a]
Ap	Alkaline phosphatase	calf	Σ
αA	α_1-Antitrypsin	human	Σ
A	Avidin	chicken	S
Bu	Butyryl cholinesterase	horse	Σ
F	Fetuin	calf	Σ
G	Glucoamylase	*Asp. niger*	B
Gl	Glucose oxidase	*Asp. niger*	S
γG[b]	γ-Globulin	bovine	S
I[c]	Invertase	yeast	B, S
Lf	Lactoferrin	bovine	Σ
Lp	Lactoperoxidase	bovine	Σ
L	Lectin	potato	Σ
αM	α_2-Macroglobulin	bovine	B
M	Mucin	bovine	S
Or	Orosomucoid	human	Σ
O	Ovalbumin	chicken	S
Ov	Ovomucoid	chicken	S
Ot	Ovotransferrin	chicken	S
P	Peroxidase	horseradish	S
Th	Thyroglobulin	porcine	Σ
Tf	Transferrin	human	Σ

[a] B: Boehringer (Mannheim); S: Serva (Heidelberg); Σ: Sigma (Munich).
[b] γG(H): H-chain; γG(L): L-chain.
[c] IB: invertase from Boehringer; IS: invertase from Serva.

bovine serum albumin (0—70 µg) or leucine (0—30 µg), respectively. Since at higher protein concentrations the method of Bradford showed deviations from Lambert-Beer's law, preferably the Moore method was used.

b) Carbohydrate determination: The determination of the carbohydrate content was performed using the phenol [23] or orcinol test [24]. Calibrations made use of mannose solutions of known concentration (0—100 or 0—300 µg). Because of its higher sensitivity, mainly the phenol test was applied.

c) Cleavage of glycoproteins: In order to cleave glycan chains from the glycoproteins, a mixture of two glycosidases (endoglycosidase F/N-glycosidase F) was used [25]. The pH optima of the two enzymes differ (5.0—7.0 and 7.0—8.0). Therefore, assays at pH 5.5 and pH 7.0 were performed; the digest at pH 5.5 was found to be more effective. Incubation with glycosidases was performed at 37°C for 20 h. Preincubation of the glycoproteins at 95°C for 3 min accelerated and completed the reaction, as a consequence of protein denaturation.

3.2. Determination of partial specific volume

a) Experimental determination: Apparent specific volumes, ϕ_2, of native proteins were determined at 4°C according to Eq. (1):

$$\phi_2 = 1/\rho_s \cdot [1 - (\rho - \rho_s)/c_2] , \tag{1}$$

where ρ is the density of a solution of definite protein concentration c_2 ($\simeq 20$ mg/ml), and ρ_s is the density of the solvent (dilute buffer). Measurements were performed in a Paar digital density meter DMA 02 [26]. Since the concentration dependence of ϕ_2 of proteins is negligible, these apparent volumes may be taken as partial specific volumes, $\bar{v}_2$. Values were converted from 4°C to room temperature (20—25°C) by adding 0.008 cm³/g [27, 28].

b) Calculation: Since experimental determinations of $\bar{v}_2$ are not always feasible, calculations of $\bar{v}_2$ from the individual components may be used instead (cf. [28]). In this context, in the case of glycoproteins, exact knowledge of the volume increments of amino acid residues and carbohydrate components and their percentage is required. In general, the values for the individual carbohydrate components are unknown. If, however, the weight fractions of protein and carbohydrate moieties are available, $\bar{v}_2$ of the native glycoprotein may be determined with sufficient accuracy ($\pm 2\%$) based on Eq. (2):

$$\bar{v}_{2,\text{calc.}} = f_P \bar{v}_P + f_{CH} \bar{v}_{CH} , \tag{2}$$

where f_P and f_{CH} are the weight fractions of the protein or carbohydrate moieties, respectively, and $\bar{v}_P$ and $\bar{v}_{CH}$

are their assumed partial specific volumes ($\bar{v}_a = 0.735$ or 0.61 cm^3/g at room temperature) [29, 30].

3.3. Nondenaturing PAGE

PAGE of native proteins was performed in the absence of denaturing agents according to Ornstein [31] and Davis [32], using a discontinuous gel system. This method of electrophoresis allows the rapid registration of impurities, fragmentation, aggregation or dissociation products of proteins.

Conditions of electrophoresis: slab gels of 2 mm thickness; for the stacking gel: acrylamide concentration $T = 4.2\%$ (cf. Eq. (4)) and tris(hydroxymethyl)aminomethane ("Tris")/HCl pH 6.8; for the resolving gel: $T = 7.7\%$ and Tris/HCl pH 9; 50 μg of protein ($c_2 = 1$ mg/ml) per slot; electrophoresis buffer: Tris/glycine pH 8.3; electrophoresis at 4°C for about 100—150 min at 40—50 mA.

3.4. SDS-PAGE

a) Basic principles and definitions:

SDS-PAGE (cf. [5—7, 13, 33—37]) was performed according to Laemmli [38, 39]. In this approach, the gels are discontinuous, both with respect to the buffer system and the acrylamide concentration. Due to the presence of the anionic detergent sodium dodecyl sulfate (SDS), the migration depends on both the molecular size of the protein subunits (including bound SDS) and their overall negative charge. The shape of the unfolded protein subunits is assumed to be identical. Therefore the method is suitable for the determination of M_r of protein subunits.

Conditions of SDS-PAGE: 2 mm slab gels; stacking gel: Tris/HCl buffer pH 6.8, (i) $T = 2.5—10\%$ (cf. Eq. (4)), $C = 2.6\%$ (cf. Eq. (5)), (ii) $T = 7.5\%$, $C = 1.3—20\%$, (iii) $T = 2.6\%$, $C = 19\%$; resolving gel: Tris/HCl buffer pH 8.7, (i) $T = 5—20\%$, $C = 1.3\%$, (ii) $T = 15\%$, $C = 0.65—10\%$, (iii) $T = 5\%$, $C = 10\%$; incubation of proteins: 100 μl protein ($c_2 = 1$ mg/ml) plus 100 μl 10 mM sodium phosphate buffer pH 7.0, containing 1% SDS and 1% 2-mercaptoethanol, incubated at 37°C for 2 h; after incubation, 10 μl 2-mercaptoethanol, 35 μl glycerol and 5 μl Bromophenol Blue (0.05%) were added; 25 μl solution ($\triangleq 10$ μg protein) per slot; electrophoresis buffer: Tris/glycine pH 8.3 containing 0.1% SDS; electrophoresis at room temperature for about 100—150 min at 40—50 mA.

Protein staining was performed by Coomassie Blue, carbohydrate staining according to [40]. Pilot tests showed that the same results concerning position, number, width, and relative intensity of the bands were obtained for both staining procedures. In the case of invertase, only a broad, diffuse band was registered with both staining methods, an effect obviously caused by heterogeneity of the carbohydrate [41].

After electrophoresis, the migration of the major band(s) was converted into relative mobilities, using Bromophenol Blue as tracking dye. This way, different runs and different types of gels could be easily normalized. The distance of migration of a protein depends on both the acrylamide concentration and the degree of crosslinking.

The following definitions hold:

$$R_f = \frac{\text{distance migrated by protein}}{\text{distance migrated by dye}} \qquad (3)$$

$$T = \frac{(m_{Ac} + m_{Bis})}{V} \cdot 100 \qquad (4)$$

$$C = \frac{m_{Bis}}{(m_{Ac} + m_{Bis})} \cdot 100 , \qquad (5)$$

where R_f is the relative electrophoretic mobility, T the acrylamide concentration (in %), C the degree of crosslinking (in %), m_{Ac} and m_{Bis} the mass (in g) of acrylamide and bisacrylamide, respectively, and V the final volume (in ml).

In some studies, T of the resolving gel is twice T of the stacking gel, and C of the stacking gel is twice C of the resolving gel, thus leading to a concentration effect (i.e., to a sharpening of the bands). For comparative studies with different T-values, however, C should be constant, since the distance of migration of a protein depends on both T and C. The choice of C is not critical; therefore, sometimes only the T value is mentioned. (If only one T and C value pair is given for an electrophoresis gel, this refers to the resolving gel).

Different sizes of $M_{r, SU}$ necessitate gels with quite different T-values (3—20%): Gels with low T-values are necessary for high molar masses, and vice versa. It should be noted that gels with very low or very high T are difficult to handle (low-T gels are easily damaged, while high-T gels tend to become brittle and opaque).

For the following experiments with proteins of varying molar subunit masses, both gel concentration and crosslinking were varied. Gels with uniform concentration rather than gradient gels were applied in order to determine correlations with respect to the electrophoretic parameters T and C.

b) Evaluation procedures and plots:

For the determination of $M_{r, SU}$ and the interpretation of data, the following evaluation methods and plots may be used:

(i) Plot of $\log M_{r, SU}$ vs. R_f (or distance of migration): By use of marker proteins (usually nonconjugated proteins), a calibration curve is obtained which can generally be approximated by a straight line. Deviations from linearity may occur especially at very low and very high R_f values. Use of the calibration profile allows the determination of $M_{r, SU}$ of substances of unknown mass. If T

and/or C were varied, such plots have to be devised for each T and each C, yielding individual calibration curves and apparent molar masses for each condition chosen.

(ii) Plot of $\log R_f$ vs. T (Ferguson plot): For a systematic variation of T (at constant C) preferably such a plot should be applied for each substance (and each C). The ordinate section and the slope of such linear graphs allow the determination of two further parameters according to:

$$\log R_f = \log Y_0 - K_R \cdot T , \qquad (6)$$

where Y_0 is the free electrophoretic mobility (describing migration in the absence of a sieving medium), and K_R is the retardation coefficient (depending on the effective molecular radius of a molecule) [13, 42, 43]. Y_0 is nearly identical for different nonconjugated proteins (since in this case all proteins bind about the same amount of SDS), but varies for different glycoproteins. K_R differs markedly both for nonconjugated proteins and glycoproteins. The identity of Y_0 values provides the theoretical justification for calculating $M_{r,SU}$ from R_f at a single T [43].

(iii) Plot of $M_{r,SU}$ vs. T: In certain cases estimates of the molar mass may be facilitated by plotting the apparent $M_{r,SU}$ against T. Anomalous behaviour becomes detectable from differences in the apparent molar mass for the same protein with varying T. While nonconjugated proteins give consistent masses at all T-values, glycoproteins are reported to have decreasing apparent masses with increasing T. According to Segrest et al. [11, 12], the asymptotic minimal molar mass for each glycoprotein at high T comes close to the real molar mass.

(iv) Plot of K_R vs. $M_{r,SU}$: In the case of nonconjugated proteins, plotting of K_R (obtained from the Ferguson plot) against $M_{r,SU}$ can be approximated by a straight line which may be used to determine the molar subunit mass, $M_{r,SU}$. In this context the question arises, whether the calibration obtained for nonconjugated proteins can also be used for the M_r determination of glycoproteins.

(v) Plot of Y_0 vs. $M_{r,SU}$ (or K_R): Similar to the above suggestions, a series of further correlation plots may be constructed (e.g., [13]). However, no significant additional information can be extracted from such plots, apart from the detection of proteins migrating anomalously on SDS gels.

(vi) Plot of R_f vs. C: Systematic variation of C (at constant T) is best represented in this plot, where each substance (and each T) yields concavely shaped curves. The usefulness of the variation of C is not well-established yet, apart from the advantages with respect to the optimization of the gels (stability, R_f-value). It seems as if highly cross-linked gels (in connection with low T) may be effectively used to estimate large molar masses [37].

c) Problems with glycoproteins:

Application of SDS-PAGE to glycoproteins poses a series of problems, which are of interest in connection with their anomalous electrophoretic behaviour:

(i) The amount of SDS binding to different glycoproteins is not constant, since SDS binds primarily to the protein moiety (cf. [15]). This generally leads to a decrease of the charge/mass ratio, thus lowering the mobility during electrophoresis and yielding too high apparent $M_{r,SU}$ estimates, especially for glycoproteins with high carbohydrate content. Usually the determination of $M_{r,SU}$ is based on calibrations with nonconjugated proteins, which contain approximately the same amount of SDS per gram of protein (about 1.4 g/g); this value, however, may vary considerably for proteins of low $M_{r,SU}$. According to Segrest and co-workers [11, 12] the errors in $M_{r,SU}$ of glycoproteins are lowered or eliminated by using gels with high T. In the light of the above considerations (concerning the relation of T and $M_{r,SU}$, cf. section 3.4.a), it may be questioned whether this approach can be valid for high-molar mass glycoproteins (which need low T).

(ii) The absolute values of the partial specific volume may influence the electrophoretic behaviour depending on the carbohydrate content and the amount of SDS binding. Since $\bar{v}_2$ of native glycoproteins varies in a linear fashion between 0.61 and 0.735 cm³/g [29, 30], in the presence of SDS $\bar{v}_2$-values of 0.61 to 0.815 cm³/g are expected, assuming SDS increments of 0 to 0.08 cm³/g ($\hat{=}0$ to 1.4 g SDS/g protein). Therefore, the use of nonglycosylated proteins ($\bar{v}_a = 0.735$ cm³/g) for calibration will lead to a systematic error which could be corrected provided glycoproteins do not deviate from the nonglycosylated reference system in an anomalous way.

(iii) Carbohydrate chains of glycoproteins branching out from the main chain of the polypeptide may lead to interactions, e.g., with the electrophoresis gel matrix. This may result in perturbations of the migration in the electric field.

3.5. Analytical ultracentrifugation

Analytical ultracentrifugation (cf. [16—18, 30, 44]) made use of a Beckman Analytical Ultracentrifuge Model E, equipped with a high sensitivity ultraviolet scanner and multiplexer system and a 10 inch recorder. Runs were performed in a six-hole rotor (AnG) at 4°C, using 12 mm double sector cells (charcoal-filled epon), and an initial concentration $c_0 = 0.5$ mg/ml.

Molar masses (in g/mol), M_2, of native proteins were determined by high-speed sedimentation equilibrium (HSSE), using the meniscus depletion technique [45]:

$$M_2 = \frac{2\,R\,T}{(1 - \bar{v}_2 \rho_s) \cdot \omega^2} \cdot \frac{d\ln c_2}{d(r^2)} , \qquad (7)$$

where ω is the angular velocity and r is the distance from the centre of rotation; solvent density ρ_s can be used instead of ρ, since at low c_2 $\rho_s \approx \rho$. In the case of heterodisperse systems (e.g., in protein solutions showing association or dissociation phenomena), evaluation leads to the weight average of M_2.

Achievement of equilibrium was accelerated by using an overspeed technique (40000 rotations per minute (rpm) to spin down the protein, then reduction of speed to the desired rpm). Various rotor speeds were applied in order to record nonideality effects. Registration of data was achieved by scanning at 280 nm; equilibrium was checked by repeated scanning about 24 h after having reached equilibrium.

Evaluation of data was performed by a computer program kindly provided by G. Böhm (University of Regensburg). M_2 was obtained from plots of logarithm of absorbance, $\ln A$, vs. square of radial distance, r^2, using linear regression analysis.

The size of $\bar{v}_2$ is a critical parameter when determining M_2. In the case of nonconjugated proteins the error in M_2 is three times the error in $\bar{v}_2$, for glycoproteins it is $\leqslant 3$ [29, 30].

4. Characteristics of used proteins

For the following experiments 21 well-characterized, commercially available glycoproteins were used, and for reference 22 nonglycosylated "simple" proteins. Proteins differed in relative molar mass (M_r = 28000—725000), subunit size ($M_{r,SU}$ = 17000—335000), isoelectric point (pI = 2—11), and carbohydrate content (2.5—51%).

After critically reviewing available data, the characteristics of the selected proteins (e.g., carbohydrate content, M_r, number of subunits (n_{SU}), $M_{r,SU}$, $\bar{v}_2$, pI, absorption coefficient) were taken from the literature [5—7, 15, 28—30, 34, 35, 41, 46—66].

5. Determination of protein and carbohydrate content of glycoproteins

Since the carbohydrate moiety is responsible for the specific behaviour of glycoproteins, the composition of the glycoproteins was analyzed with respect to their protein and carbohydrate content.

Table 3. Experimental and calculated partial specific volumes[a]) of glycoproteins, arranged according to increasing carbohydrate content

Protein	Literature		Densimetry	Calculation
	CH (%)	$\bar{v}_{2,\text{exp.}}$ (cm³/g)	$\bar{v}_{2,\text{exp.}}$ (cm³/g)	$\bar{v}_{2,\text{calc.}}$ (cm³/g)
γ-Globulin	2.5	0.730		0.732
Ovotransferrin	2.5	0.732	0.729 ± 0.000	0.732
Ovalbumin	3.0	0.748	0.726 ± 0.000	0.731
Transferrin	5.9	0.719	0.714 ± 0.001	0.728
Lactoferrin	7.7			0.725
Thyroglobulin	8.5		0.724 ± 0.006	0.724
α₂-Macroglobulin	9.0	0.734		0.724
Alkaline phosphatase	9.7[b])		0.718 ± 0.005	0.723
α₁-Antitrypsin	12.4			0.720
Glucose oxidase	16.5	0.711	0.712 ± 0.002	0.714
Peroxidase	16.5	0.699		0.714
Lactoperoxidase	18.3[b])			0.712
Avidin	19.0			0.711
Butyryl cholinesterase	20.0	0.707		0.710
Glucoamylase	22.0			0.708
Fetuin	23.0	0.702	0.704 ± 0.003	0.706
Ovomucoid	23.0	0.693		0.706
Orosomucoid	42.0	0.689		0.683
Mucin	44.0			0.680
Lectin	50.0			0.673
Invertase	51.0		0.652 ± 0.001[c])	0.671

[a]) All values are given for room temperature.
[b]) Value determined in this study.
[c]) Value determined for invertase from Boehringer.

The results agreed within ±1% with literature data (Table 3).

6. Determination of partial specific volume

The partial specific volumes of native nonconjugated proteins are well known [28]. In the case of glycoproteins, previous findings that $\bar{v}_2$ of native glycoproteins decreases linearly with increasing carbohydrate content [29, 30], were confirmed. As shown in Table 3, there is good agreement between present data and data in the literature. The table also contains $\bar{v}_2$-values calculated on the basis of the known carbohydrate content; for sake of consistency, these calculated values were used in the following studies.

7. SDS-PAGE

SDS-PAGEs of both nonconjugated proteins and glycoproteins were performed varying both T and C separately.

First, nonconjugated proteins were investigated by varying T (7.5 to 20% for the resolving gel, in steps of 2.5%). R_f-values were derived, and calibration curves established for each T. Then the same approach was applied to glycoproteins, using the calibration lines derived for nonconjugated proteins, in order to estimate the relative molar subunit masses of glycoproteins. Finally, C was varied (0.65 to 10% for the resolving gel), in order to investigate the influence of crosslinking. To find out possible correlations, different evaluation procedures were used and results were presented in suitable plots.

7.1. Variation of acrylamide concentration

The electrophoreses of the nonconjugated proteins gave only straight lines in definite ranges of R_f, depending on the masses of the proteins and the T-values used. Two typical examples ($T = 7.5$ and 20%) are shown in Fig. 1 which demonstrates that most proteins ($M_{r,SU} > 30000$) are located on a straight line (especially apoferritin, phosphorylase, catalase, lactate dehydrogenase, glyceraldehyde-3-phosphate dehydrogenase); deviations are reproducible. At $T < 10\%$, proteins with $M_{r,SU} < 30000$ are below the regression line (Fig. 1A), whereas at $T > 10\%$ they coincide or lie slightly above the line (Fig. 1B).

Fig. 2A shows the calibration lines, obtained for different T-values for nonconjugated proteins. Obviously there is a linear correlation of both the intercepts on the ordinate and the slopes vs. T (Fig. 2B). The ordinate sections (which reflect the exclusion limit of gels) vary only slightly due to the logarithmic scale.

In contrast to the "simple" (nonglycosylated) proteins, the electrophoretic patterns of the glycoproteins deviate from the linear relationship in $\log M_{r,SU}$ vs. R_f plots (Fig. 3). Anomalies hold especially for proteins with high $M_{r,SU}$ at high T (cf. the values for α_2-macroglobulin and thyroglobulin). Using the calibration lines obtained for the nonconjugated proteins to estimate apparent molar masses of glycoproteins, positive and negative deviations (i.e. too high and too low apparent molar masses) are obtained.

The results for all glycoproteins and all T-values investigated in the present study are summarized in Table 4. $M_{r,SU}$-values showing the best agreement with data reported in the literature are italicized. In general the accordance is better than 10%, in some cases large deviations occur. The best results were obtained for relative molar masses between 30000 and 80000. Obviously there is no clear correlation between T-values on one hand, and the size of $M_{r,SU}$ or the carbohydrate content on the other. From the results it becomes clear that variation of T is of lesser importance compared to the correct choice of the proper range of T [5]. Using K_R (which includes the information obtained at all T-values), does not improve the result (cf. Table 4). As reported [13], in the case of anomalous migration, K_R cannot be interpreted simply in terms of molar masses.

A systematic analysis of SDS-PAGEs by Ferguson plots was performed for both nonconjugated proteins and glycoproteins. For each substance, $\log R_f$ shows a linear dependence on the T-value. As may be expected, the Y_0-values differ significantly only for glycoproteins, but not for nonconjugated proteins, whereas the K_R-values differ for both types of proteins (Fig. 4).

Trying to combine all available parameters and data for glycoproteins (e.g., by plotting Y_0 vs. K_R or $M_{r,SU}$ or the carbohydrate content etc.) no clear correlation could be detected [67]. Figure 5 illustrates this by plotting the Y_0 and K_R dependence on the carbohydrate content for the whole sample of glycoproteins. Obviously, there is no reliable way to determine the molar mass of glycoproteins by SDS-PAGE.

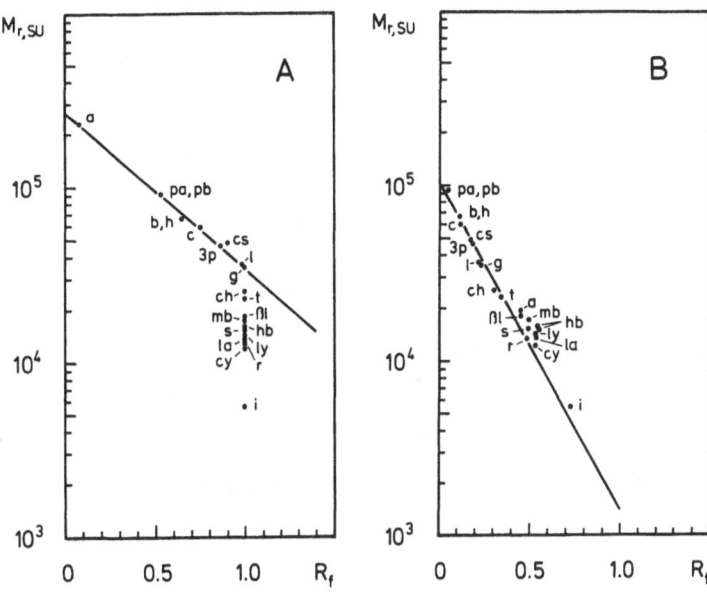

Fig. 1. SDS-PAGE of nonconjugated proteins at variable T (A: 7.5%, B: 20%) and constant C (1.3%), presented in a plot of logarithm of relative molar subunit mass, $\log M_{r,SU}$, vs. relative mobility, R_f. Symbols as explained in Table 1. Under the given conditions (incubation with 1% SDS), (apo)ferritin dissociates into the dodecameric half molecule and/or the subunit

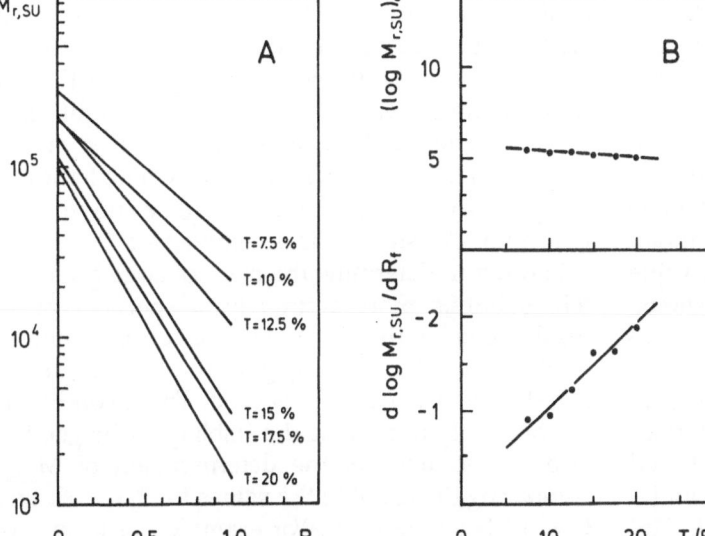

Fig. 2. SDS-PAGE of nonconjugated proteins. A: Summary of calibration lines obtained at variable T (7.5—20%) and constant C (1.3%) in a plot of $\log M_{r,SU}$ vs. R_f. B: Ordinate sections (top) and slopes (bottom), obtained from the calibration lines in A

There are numerous possible reasons for the imperfect correlation of the apparent molar mass of glycoproteins and their carbohydrate content: carbohydrate heterogeneity, volume changes, interactions with the gel etc. (i) Significant heterogeneity of the carbohydrate content can be excluded in most cases, based on the analysis of the carbohydrate contents and AUC (different masses at different rpm: see below). (ii) Alterations of the hydrodynamic volume may be attributed to changes in SDS binding, $\bar{v}_2$ and/or hydration. No systematic effect of the carbohydrate content on the apparent molar mass was observed. (iii) The conformation of the glycan chains of glycoproteins differs markedly, thereby giving rise to a strong influence on the behaviour during SDS-PAGE. Interactions between the carbohydrate moiety and the electrophoresis gel are expected to be complex, since the carbohydrates, in contrast to the polypeptide chain, are not solubilized by SDS. As a consequence, tem-

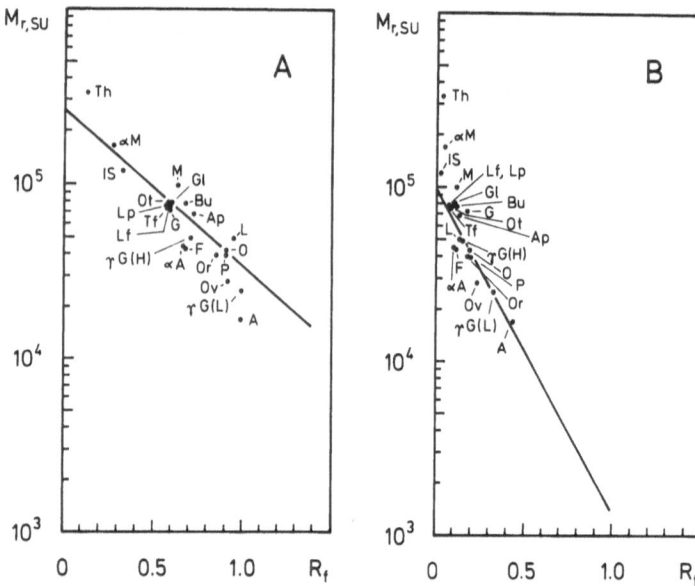

Fig. 3. SDS-PAGE of glycoproteins at variable T (A: 7.5%, B: 20%) and constant C (1.3%), presented in a plot of $\log M_{r,SU}$ vs. R_f. The calibration lines were taken from the corresponding plots for nonconjugated proteins. Symbols as in Table 2

porary entanglement of the antennae with the gel matrix or lubrication effects may be envisaged which may alter the migration in the gel, and, hence, the apparent molar mass.

Plotting the apparent molar mass of glycoprotein subunits vs. T stresses the peculiar behaviour of certain glycoproteins (Fig. 6). As taken from Table 4, neither the smallest value for the apparent molar mass (not always found at highest T) nor the value at highest T reflects the true $M_{r,SU}$-value. Obviously, the suggestion [11, 12] that the smallest mass (obtained at maximum T) represents the true molar mass needs to be corrected, especially for glycoproteins of high $M_{r,SU}$. In this context, it is important to note that Segrest's recommendation was based on only a few glycoproteins of rather low molar mass ($M_{r,SU} \simeq 20000-60000$). Leach et al. [15] reported that the apparent $M_{r,SU}$ not always approaches the true molar mass in an asymptotic fashion; instead each glycoprotein may exhibit unique properties. This observation is in accord with the present analysis (cf. Table 4).

7.2. Variation of degree of crosslinking

The results of the variation of the degree of crosslinking (C) at definite T-value, may be presented in R_f vs. C plots. For both nonconjugated proteins and glycoproteins concave curves of

similar shape are obtained (Fig. 7). Only in a few cases slight crossing of the curves is observed, generally only at extremes of C. This means that, as long as the shape of the curves varies uniformly with C, similar $M_{r,SU}$-values are obtained at different C-values. Therefore, in cases where handling the gels may cause problems, C can be chosen such that optimum stability is accomplished.

In order to determine the mass of large glycoprotein subunits more accurately, a special type of resolving gel with low acrylamide concentration ($T = 5\%$, suited for high $M_{r,SU}$) and a high degree of crosslinking ($C = 10\%$) was used. These conditions not only help improving the stability of the gel, but also the accuracy of the determination of $M_{r,SU}$, comparing the result to the values for $T = 7.5\%$ and $C = 1.3\%$ (cf. Table 4). For example, for α_2-macroglobulin and thyroglobulin $M_{r,SU} = 185000$ and 248000 are obtained, as compared to $M_{r,SU} = 170000$ and 335000, reported in the literature.

7.3. Removal of the carbohydrate moiety of glycoproteins

In the present study, invertase from yeast represents the glycoprotein with the highest carbohydrate content (51%). Using glycosidases to remove most of the carbohydrate, the electrophoretic behaviour of the glycoprotein and the carbohydrate-depleted form could be compared.

Table 4. Relative molar subunit masses of glycoproteins as obtained from SDS-PAGE (T = 7.5—20%, C = 1.3%), arranged according to increasing carbohydrate content

| Protein | Literature | | SDS-PAGE[a,b] | | | | | | |
	CH (%)	$M_{r,SU} \cdot 10^{-3}$	T=7.5% $M_{r,SU} \cdot 10^{-3}$	T=10.0% $M_{r,SU} \cdot 10^{-3}$	T=12.5% $M_{r,SU} \cdot 10^{-3}$	T=15.0% $M_{r,SU} \cdot 10^{-3}$	T=17.5% $M_{r,SU} \cdot 10^{-3}$	T=20.0% $M_{r,SU} \cdot 10^{-3}$	via K_R $M_{r,SU} \cdot 10^{-3}$
γ-Globulin(L)	0	25.0	35.0	31.0	30.0	24.0	27.0	*25.0*	32.0
Ovotransferrin	2.5	78.0	*81.0*	*81.0*	83.0	73.0	*75.0*	71.0	62.5
Ovalbumin	3.0	43.0	*42.5*	45.0	50.0	41.5	40.0	45.0	44.0
γ-Globulin(H)[c]	3.8	50.0	64.0	62.0	68.0	58.0	58.0	*54.0*	56.7
Transferrin	5.9	76.0	82.0	81.0	81.5	72.5	*74.5*	73.0	62.5
Lactoferrin	7.7	77.0	82.0	80.0	94.0	*77.0*	75.0	72.0	66.0
Thyroglobulin	8.5	335.0	*210.0[d]*	150.0	155.0	140.0	110.0	94.0	126.0
α₂-Macroglobulin	9.0	170.0	*155.0[d]*	130.0	145.0	120.0	100.0	80.0	86.0
Alkaline phosphatase	9.7[e]	69.0	62.0	70.0	*69.0*	*69.0*	58.0	60.0	57.5
α₁-Antitrypsin	12.4	45.0	70.0	70.0	84.0	68.0	*58.0*	64.0	58.5
Glucose oxidase	16.5	80.0	*80.0*	80.5	93.0	76.0	73.0	67.0	69.0
Peroxidase	16.5	40.0	42.5	45.0	50.0	41.0	*39.5*	44.0	44.8
Lactoperoxidase	18.3[e]	77.5	81.0	*80.0*	81.5	71.0	74.0	72.0	61.5
Avidin	19.0	17.0	35.5	22.0	*17.0*	13.0	12.0	15.0	17.5
Butyryl cholinesterase	20.0	79.0	69.0	70.0	66.0	*70.0*	62.0	64.0	57.5
Glucoamylase	22.0	72.0	81.0	78.0	75.0	62.0	61.0	50.0	40.5
Fetuin	23.0	44.0	67.0	70.0	83.0	68.0	64.0	*59.0*	65.0
Ovomucoid	23.0	28.0	42.0	42.0	46.0	*35.0*	36.0	37.0	33.5
Orosomucoid	42.0	40.0	47.0	46.5	53.0	44.0	*41.0*	46.0	43.3
Mucin	44.0	100.0	74.0	78.0	*82.0*	70.0	70.0	62.0	60.0
Lectin	50.0	50.0	39.0	44.0	47.0	40.0	39.0	56.0	42.5
Invertase (IS)	51.0	120.0	140.0	*115.0*	*115.0*	90.0	83.0	90.0	77.0

[a] Accuracy: $\leq \pm 5\%$.
[b] Best agreement between literature value and experimentally determined value (T = 7.5—20%) is italicized.
[c] Carbohydrates only at H-chain.
[d] Use of very low T-values (3—5%) and/or of nonlinear calibration lines lead to a further increase of $M_{r,SU}$.
[e] Value determined in this study.

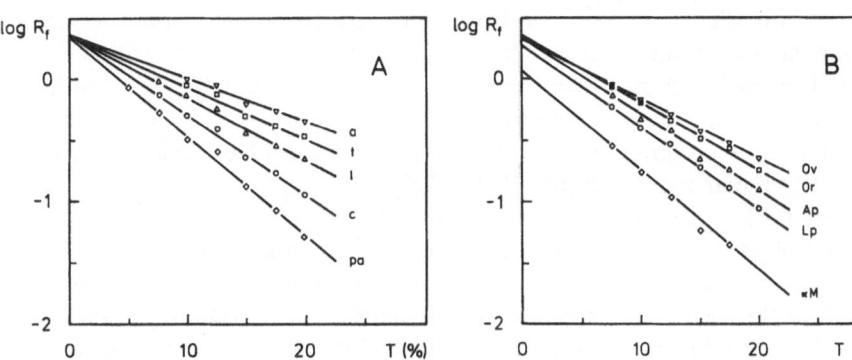

Fig. 4. Ferguson plot for selected nonconjugated and conjugated proteins. Symbols as in Tables 1 and 2.
A: Nonconjugated proteins.
B: Glycoproteins

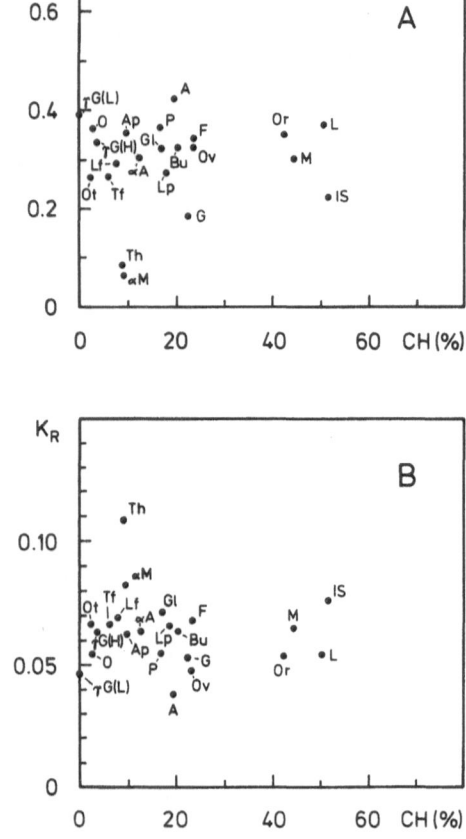

Fig. 5. Logarithm of free electrophoretic mobility of glycoproteins, $\log Y_0$ (A), or retardation coefficient, K_R (B), plotted vs. carbohydrate content, CH. Symbols as in Table 2

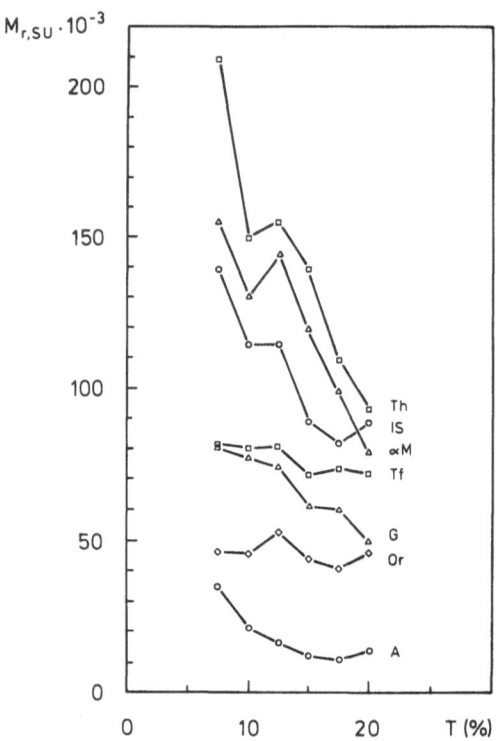

Fig. 6. SDS-PAGE of selected glycoproteins at variable T (7.5—20%) and constant C (1.3%). Plot of apparent values for relative molar subunit mass vs. acrylamide concentration, T. Symbols as in Table 2

SDS-PAGE of the glycoprotein (stacking gel: $T = 7.5\%$, $C = 2.6\%$; resolving gel: $T = 15\%$, $C = 1.3\%$) and subsequent protein and carbohydrate staining yielded a diffuse band ($M_{r,SU} \simeq 125\,000$), the broadness of which was obviously caused by heterogeneity of the carbohydrate content [41]. The same result was obtained for the digest at pH 7, which was obviously incomplete. The digest at pH 5.5, however, resulted in one sharp protein band ($M_{r,SU} = 60\,000$) indicating that the protein under this condition has lost most of its carbohydrate moiety. The agreement of the protein content of the enzyme (49%) and the molar mass observed after carbohydrate digest (48%) proves that in the case of invertase the carbohydrate content does not affect the electrophoretic behaviour in a significant way. This may be due to the fact that yeast invertase does not contain sialic acid [68], the

presence of which is suspected to be one of the factors responsible for the anomalous electrophoretic behaviour of glycoproteins [15].

8. Analytical ultracentrifugation

Results of high-speed sedimentation equilibrium experiments with the native (dialyzed) glycoproteins are presented in Table 5; a typical example of a linearized graph is shown in Fig. 8. For sake of comparison with electrophoretic data obtained from SDS-PAGE, it might be considered appropriate to work under denaturing conditions (SDS, guanidinium chloride). However, in this case, changes of $\bar{v}_2$ may lead to serious perturbations (cf. [27, 28]).

Provided correct values for the partial specific volumes are applied, for most proteins the agreement of the numerical results at different rotor speeds, on one hand, and M_r-data reported in the literature, on the other, is found to be satisfactory.

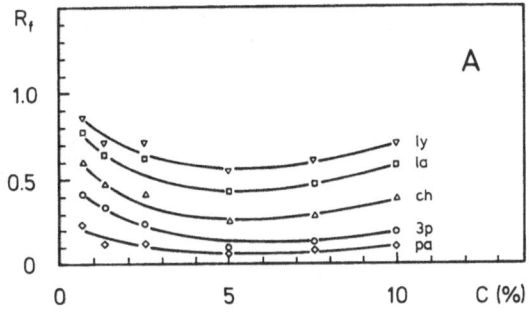

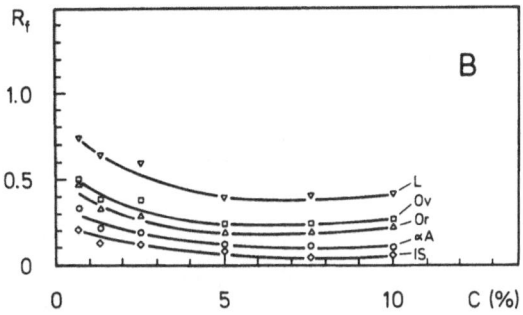

Fig. 7. SDS-PAGE of selected nonconjugated and conjugated proteins at constant T (15%) and variable C (0.65—10%). Plot of electrophoretic mobility, R_f, vs. degree of crosslinking, C. Symbols as in Tables 1 and 2. A: Nonconjugated proteins. B: Glycoproteins

This holds especially for single-chain proteins with $M_r < 100000$ (exception: α_1-antitrypsin). With higher M_r and more complicated subunit composition, deviations may occur (alkaline phosphatase, butyryl cholinesterase, lectin, mucin, and, to some extent, invertase and α_2-macroglobulin). However, these peculiarities have biochemical rather than technical or methodical reasons; these include (partial) dissociation into subunits, heterogeneity, or other nonideality effects.

An inspection of Table 5 reveals that obviously alkaline phosphatase and lectin are dissociated completely into subunits (identical low M_r at low and high rpm). Butyryl cholinesterase and mucin seem to contain a mixture of subunits and oligomeric particles (higher M_r at low rpm). Different M_r-values have also been found at low and high rpm for α_2-macroglobulin and invertase; as mentioned, this may be caused by dissociation into subunits, impurities and/or heterogeneity of the carbohydrate moiety etc. Anomalies of this kind require additional experiments apart from AUC. For example, impurities become visible in the elec-

tropherograms of native PAGEs (e.g., alkaline phosphatase, α_1-antitrypsin, butyryl cholinesterase, lectin, α_2-macroglobulin).

9. Comparison of results obtained by SDS-PAGE and analytical ultra-centrifugation

In comparing the results summarized in the previous sections, Fig. 9 clearly proves the superiority of AUC compared to SDS-PAGE. Over the whole range of carbohydrate contents and molar subunit masses, ultracentrifugation yields correct results for glycoproteins, provided the samples are pure and the partial specific volume is known. Ultracentrifugation of native proteins has to be performed such that dissociation/association phenomena become detectable (e.g., by variation of rotor speed, initial protein concentration, or wavelength for the scanning procedure). If, on the other hand, it is intended to determine the particle weight of stable assemblies, stabilizing conditions have to be chosen where subunit dissociation becomes negligible.

The application of SDS-PAGE to glycoproteins is connected with a number of problems. Even if all precautions are taken into consideration, less accurate results and atypical behaviour are obtained. Especially for glycoproteins with high molar subunit mass, perturbations may lead to much too low $M_{r,SU}$-values. There is no way to judge the correctness of the result, except by independent methods not involving matrix interactions.

Obviously, differences in both the conformation of the sugar moieties, and the interactions of the carbohydrate chains with the gel, as well as abnormal SDS binding are responsible for the anomalies in the electrophoretic mobility of glycoproteins. There is no obvious correlation between the carbohydrate content and the anomalies found for the apparent $M_{r,SU}$. Evidently, the previously mentioned perturbations are more important than the carbohydrate content. The main advantages of the SDS-PAGE are threefold: (i) small amounts of impurities do not influence the results, (ii) experiments can be performed fast, and (iii) knowledge of $\bar{v}_2$ is not necessarily required.

Molar mass estimates of glycoproteins based on SDS-PAGE should be regarded as representing only estimates. If one is satisfied with an accuracy of no more than $\pm 20\%$, SDS-PAGE is the method of

Table 5. Relative molar masses of glycoproteins derived from AUC (HSSE), arranged according to increasing M_r[a]

Protein	Literature			AUC		
	CH (%)	M_r $\cdot 10^{-3}$	n_{SU}	rpm $\cdot 10^{-3}$	M_r $\cdot 10^{-3}$	$M_{r,SU}$[b] $\cdot 10^{-3}$
Ovomucoid	23.0	28.0	1	16	28.6 ± 1.6	28.8
				20	28.9 ± 0.6	
Orosomucoid	42.0	40.0	1	16	42.3 ± 4.4	40.1
				20	37.8 ± 2.6	
Peroxidase	16.5	40.0	1	16	42.5 ± 2.8	41.9
				20	41.3 ± 3.0	
Ovalbumin	3.0	43.0	1	16	41.6 ± 2.9	42.1
				20	42.5 ± 3.4	
Fetuin	23.0	44.0	1	16	45.1 ± 5.9	44.8
				20	44.4 ± 6.7	
α_1-Antitrypsin	12.4	45.0	1	14	63.6 ± 6.3	61.1
				18	58.6 ± 6.2	
Avidin	19.0	68.3	4	14	68.9 ± 3.6	17.2
				18	68.9 ± 5.5	
Glucoamylase	22.0	72.0	1	14	72.9 ± 3.3	71.9
				18	70.9 ± 4.6	
Transferrin	5.9	76.0	1	14	85.4 ± 4.4	83.5
				18	81.5 ± 4.4	
Lactoferrin	7.7	77.0	1	12	89.6 ± 6.0	83.2
				16	76.7 ± 8.8	
Lactoperoxidase	18.3[c]	77.5	1	14	75.5 ± 4.2	75.7
				18	75.9 ± 3.5	
Ovotransferrin	2.5	78.0	1	12	87.1 ± 5.0	83.7
				16	80.3 ± 7.4	
Lectin	50.0	100.0	2	12	50.6 ± 3.4	48.3[d]
				16	46.0 ± 5.6	
Alkaline phosphatase	9.7[c]	140.0	2	12	85.2 ± 4.5	75.4[d]
				16	65.5 ± 6.8	
γ-Globulin	2.5	150.0	4	12	149.0 ± 9.5	36.3
				16	141.6 ± 11.1	
Glucose oxidase	16.5	160.0	2	10	150.4 ± 8.4	78.1
				14	161.9 ± 12.9	
Butyryl cholinesterase	20.0	317.0	4	10	149.1 ± 41.5	74.6[e]
				14	71.2 ± 3.6	
Invertase (IS)	51.0	240.0	2	4	262.3 ± 36.3	131.2[e]
				10	189.1 ± 18.1	
				14	229.7 ± 19.4	
Invertase (IB)	51.0	240.0	2	4	228.7 ± 23.6	114.4[e]
				10	155.4 ± 6.4	
				14	153.0 ± 9.2	
Mucin	44.0	500.0	5	4	453.8 ± 83.6	90.8[e]
				10	120.0 ± 20.9	
				14	75.2 ± 9.5	
Thyroglobulin	8.5	669.0	2	4	686.1 ± 70.2	332.7
				6	644.8 ± 31.3	
α_2-Macroglobulin	9.0	725.0	4	4	722.6 ± 54.7	180.7[e]
				6	465.0 ± 35.3	

[a] For sake of comparison with data from SDS-PAGE, all masses are given as relative molar masses.
[b] The $M_{r,SU}$ values have been derived from the mean values of M_r obtained from runs at different rpm by dividing by n_{SU}, except otherwise stated.
[c] Value determined in this study.
[d] Mean value of the already dissociated protein.
[e] Value obtained at lowest rpm divided by n_{SU}; in the case of butyryl cholinesterase partial dissociation is taken into account.

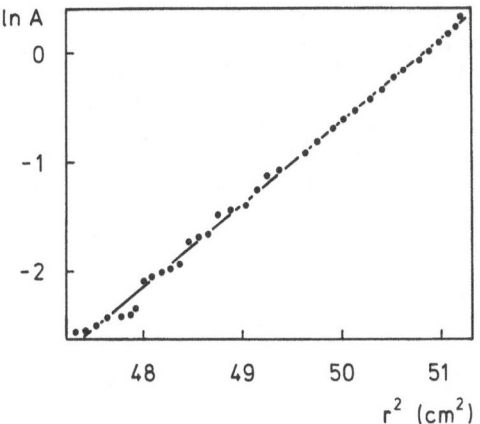

Fig. 8. Analytical ultracentrifugation (high-speed sedimentation equilibrium) of native α_2-macroglobulin (4000 rpm, 4°C). Plot of logarithm of absorbance, $\ln A$, vs. square of radial distance, r^2

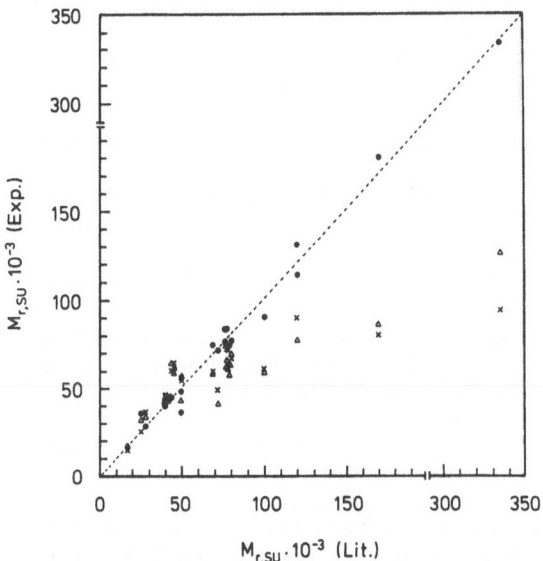

Fig. 9. Comparison of experimental and literature values for relative molar subunit masses of glycoproteins. (●): AUC, (△): SDS-PAGE, determination via K_R, (×): SDS-PAGE, determination by use of T = 20%. The line represents the median

choice. In this case, instead of varying T (and extrapolation of $M_{r,SU}$ to high T-values), measurement at one T is commonly sufficient as a first approximation. The proper choice of the correct T with respect to $M_{r,SU}$ seems to be of more importance than the variation of T. Varying T, on the other

hand, may indicate abnormal migration during electrophoresis and may reduce or suppress the error in $M_{r,SU}$ in some cases. Ultimately, the physical properties of the individual glycoproteins govern the accuracy of the estimated molar mass. At present, no reliable way to correct the anomalies of glycoproteins can be suggested. Aberrant mobilities during SDS-PAGE cannot be correlated with the amount of carbohydrate. Thus, for accurate determinations of molar masses more rigorous procedures such as sedimentation equilibrium in the analytical ultracentrifuge have to be employed.

Acknowledgements

We are much indebted to G. Kern and L. Lehle for helpful discussions, and to Mrs. S. Richter for skilful technical assistance. Work was supported by the Deutsche Forschungsgemeinschaft and the Fonds der Chemischen Industrie.

References

1. Gottschalk A, ed (1972) Glycoproteins, Parts A and B, 2nd ed. Elsevier, Amsterdam — London — New York
2. Gibbons RA (1972) In: Gottschalk A (ed) Glycoproteins, Part A, 2nd ed. Elsevier, Amsterdam — London — New York, pp 31—140
3. Fish WW (1975) In: Korn ED (ed) Methods in Membrane Biology, Vol 4. Plenum Press, New York — London, pp 189—276
4. Beeley JG (1985) In: Burdon RH, van Knippenberg PH (eds) Laboratory Techniques in Biochemistry and Molecular Biology, Vol 16. Elsevier, Amsterdam — New York — Oxford, pp 63—99
5. Andrews AT (1986) Electrophoresis: Theory, Techniques, and Biochemical and Clinical Applications, 2nd ed. Clarendon Press, Oxford
6. Wagner H, Blasius E (eds) (1989) Praxis der elektrophoretischen Trennmethoden. Springer-Verlag, Berlin — Heidelberg — New York — London — Paris — Tokyo
7. Hames BD, Rickwood D (eds) (1990) Gel Electrophoresis of Proteins, A Practical Approach, 2nd ed. IRL Press at Oxford University Press, Oxford — New York — Tokyo
8. Schubert D (1970) J Mol Biol 51:287—301
9. Bretscher MS (1971) Nature New Biol 231:229—232
10. Glossmann H, Neville DM Jr (1971) J Biol Chem 246:6339—6346
11. Segrest JP, Jackson RL, Andrews EP, Marchesi VT (1971) Biochem Biophys Res Commun 44:390—395
12. Segrest JP, Jackson RL (1972) Methods Enzymol 28:54—63
13. Banker GA, Cotman CW (1972) J Biol Chem 247:5856—5861

14. Westphal U, Burton RM, Harding GB (1975) Methods Enzymol 36:91—104
15. Leach BS, Collawn JF Jr, Fish WW (1980) Biochemistry 19:5734—5741
16. Creeth JM, Pain RH (1967) Progr Biophys Molec Biol 17:217—287
17. Chervenka CH (1973) A Manual of Methods for the Analytical Ultracentrifuge. Spinco Division of Beckman Instruments, Palo Alto
18. Van Holde KE (1975) In: Neurath H, Hill RC (eds) The Proteins, Vol 1, 3rd ed. Academic Press, New York — San Francisco — London, pp 225—291
19. Kirschner K, Voigt B (1968) Hoppe-Seyler's Z Physiol Chem 349:632—644
20. Stallcup WB, Mockrin SC, Koshland DE Jr (1972) J Biol Chem 247:6277—6279
21. Bradford MM (1976) Anal Biochem 72:248—254
22. Moore S (1968) J Biol Chem 243:6281—6283
23. Dubois M, Gilles KA, Hamilton JK, Rebers PA, Smith F (1956) Anal Chem 28:350—356
24. François C, Marshall RD, Neuberger A (1962) Biochem J 83:335—341
25. Boehringer (1987) Glycohydrolases. Boehringer, Mannheim
26. Kratky O, Leopold H, Stabinger H (1973) Methods Enzymol 27:98—110
27. Durchschlag H, Jaenicke R (1982) Biochem Biophys Res Commun 108:1074—1079
28. Durchschlag H (1986) In: Hinz H-J (ed) Thermodynamic Data for Biochemistry and Biotechnology. Springer Verlag, Berlin — Heidelberg — New York — Tokyo, pp 45—128
29. Durchschlag H (1988) Biochem (Life Sci Adv) 7:181—188
30. Durchschlag H (1989) Colloid Polym Sci 267:1139—1150
31. Ornstein L (1964) Ann N Y Acad Sci 121:321—349
32. Davis BJ (1964) Ann N Y Acad Sci 121:404—427
33. Shapiro AL, Viñuela E, Maizel JV Jr (1967) Biochem Biophys Res Commun 28:815—820
34. Dunker AK, Rueckert RR (1969) J Biol Chem 244:5074—5080
35. Weber K, Osborn M (1969) J Biol Chem 244:4406—4412
36. Weber K, Pringle JR, Osborn M (1972) Methods Enzymol 26:3—27
37. Kondo T, Tarutani O, Ui N (1981) J Biochem 89:379—384
38. Laemmli UK (1970) Nature 227:680—685
39. Laemmli UK, Favre M (1973) J Mol Biol 80:575—599
40. Zacharius RM, Zell TE, Morrison JH, Woodlock JJ (1969) Anal Biochem 30:148—152
41. Colonna WJ, Cano FR, Lampen JO (1975) Biochim Biophys Acta 386:293—300
42. Ferguson KA (1964) Metab Clin Exp 13:985—1002
43. Neville DM Jr (1971) J Biol Chem 246:6328—6334
44. Durchschlag H, Jaenicke R (1983) Int J Biol Macromol 5:143—148
45. Yphantis DA (1964) Biochemistry 3:297—317
46. Smith MH (1970) In: Sober HA (ed) CRC Handbook of Biochemistry, 2nd ed. The Chemical Rubber Co, Cleveland, pp C3—C35
47. Klotz IM, Darnall DW (1970) In: Sober HA (ed) CRC Handbook of Biochemistry, 2nd ed. The Chemical Rubber Co, Cleveland, pp C62—C66
48. Fasman GD (ed) (1976) CRC Handbook of Biochemistry and Molecular Biology, 3rd ed, Proteins Vol 2 and 3. CRC Press, Cleveland
49. Soodak M (1976) In: Fasman GD (ed) CRC Handbook of Biochemistry and Molecular Biology, 3rd ed, Proteins Vol 2. CRC Press, Cleveland, pp 257—275
50. Dayhoff MO (ed) (1972) Atlas of Protein Sequence and Structure, Vol 5; (1973) Vol 5 Suppl 1; (1976) Vol 5 Suppl 2; (1978) Vol 5 Suppl 3. National Biomed Research Foundation, Washington DC
51. Schultze HE, Heremans JF (1966) Molecular Biology of Human Proteins: With Special Reference to Plasma Proteins, Vol 1. Elsevier, Amsterdam — London — New York, pp 173—235
52. Jakubke H-D, Jeschkeit H (eds) (1981) Lexikon Biochemie, 2nd ed. Verlag Chemie, Weinheim
53. Leach BS, Collawn JF Jr, Fish WW (1980) Biochemistry 19:5741—5747
54. Kempner ES, Miller JH, McCreery MJ (1986) Anal Biochem 156:140-146
55. Peters T Jr (1985) Adv Prot Chem 37:161—245
56. Trimble RB, Maley F (1977) J Biol Chem 252:4409—4412
57. Malamud D, Drysdale JW (1978) Anal Biochem 86:620—647
58. Treffry A, Harrison PM (1979) Biochem J 181:709—716
59. Fischbach FA, Anderegg JW (1965) J Mol Biol 14:458—473
60. Crichton RR, Charloteaux-Wauters M (1987) Eur J Biochem 164:485—506
61. Svensson B, Pedersen TG, Svendsen I, Sakai T, Ottesen M (1982) Carlsberg Res Commun 47:55—69
62. Clarke AJ, Svensson B (1984) Carlsberg Res Commun 49:559—566
63. Kronman MJ, Andreotti RE (1964) Biochemistry 3:1145—1151
64. Andrews P (1965) Biochem J 96:595—606
65. Bloomfield VA (1983) Biopolymers 22:2141—2154
66. Teng T-L, Harpst JA, Lee JC, Zinn A, Carlson DM (1976) Arch Biochem Biophys 176:71—81
67. Christl P (1990) Diplomarbeit. University of Regensburg, FRG
68. Neumann NP, Lampen JO (1967) Biochemistry 6:468—475

Authors' address:

Dr. Helmut Durchschlag
Institut für Biophysik und Physikalische Biochemie
Universität Regensburg
Universitätsstraße 31
D-8400 Regensburg, FRG

Progress in Colloid & Polymer Science

Progr Colloid Polym Sci 86:57—61 (1991)

Analytical ultracentrifuges with multiplexer and video systems for measuring particle size and molecular mass distributions

B. Ortlepp and D. Panke

Röhm GmbH, Darmstadt, FRG

Abstract: The significance of the analytical ultracentrifuge for industrial research has certainly not decreased; it is still an important instrument for controling the production of aqueous as well as organic emulsion polymers, and for the development and analysis of macromolecules. In order to meet increasing demands with respect to the number of samples and of research problems the use of multiple place rotors and of multiplexer systems has become necessary. The use of a homemade multiplexer system for measuring particle size distributions will be demonstrated by some examples. Furthermore, the conventional Schlieren camera was replaced by a pulsed video camera. Thereby, the simultaneous detection of the Schlieren images of up to six polymer solutions has become possible. Controlling the camera and evaluating the video images is done with a personal computer, and the MMD is calculated on a VAX host computer. Finally, a small but significant improvement of the UV system of the Beckman E allows for more reliable scanner measurements with synthetic polymers, as was successfully tested with PMMA samples. For measurements of MMDs the UV system may replace or complement the Schlieren optics.

Key words: Analytical ultracentrifuge; particle size distribution; molecular mass distribution; digital image processing; UV scanner

Introduction

The significance of the analytical ultracentrifuge (AUC) for industrial research has not decreased during the last decades, but — after a temporary standstill — is rather increasing. In our company the ultracentrifuge has an unchallenged importance for the determination of particle size distributions of emulsion polymers and for the measurement of molecular masses and their distributions of those polymers where light scattering and size exclusion chromatography fail.

In order to speed up these principally quite lengthy procedures, it is necessary to work with multiple place rotors and with multiplexers and to employ electronic data acquisition and processing.

The main applications of the two analytical ultracentrifuges (Beckman model E) in our research department are as follows: standard sedimentation velocity runs for measuring sedimentation coefficient S and mean molecular mass M_s; molecular mass distribution (MMD) of poly(meth)acrylates with very high molecular mass and of cationic polyelectrolytes [1]; particle size distribution (PSD) of aqueous and of organic latexes; density gradients.

The video Schlieren optics are used for measuring sedimentation velocities, density gradients, and MMDs of unpolar as well as of charged polymers. PSDs are determined via turbidity measurements according to Scholtan and Lange [2] in multiplexer systems with six place rotors. With the UV optics we aim at getting reliable MMDs at concentrations as small as possible.

Experimental

Figure 1 shows, for the first machine, the scheme of the Schlieren optics (with pulsed video camera) and that of both the standard scanner and the turbitity measurement

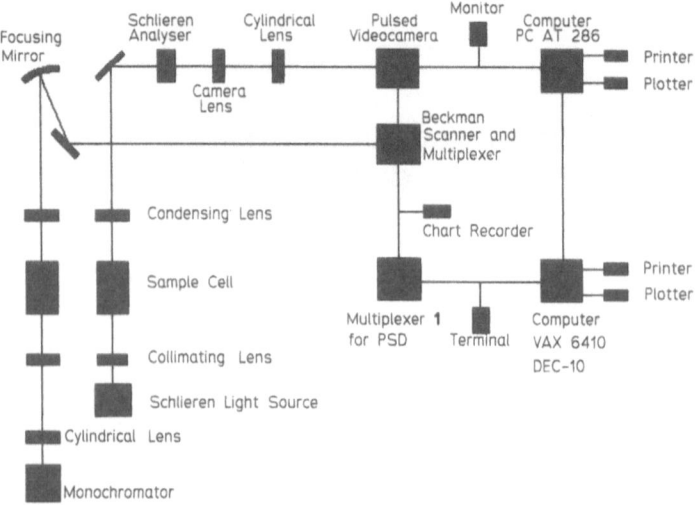

Fig. 1. Optical pathways of the first AUC system with video Schlieren optics (multiplexer method), multiplexer system for PSD, and UV photoelectrical scanner

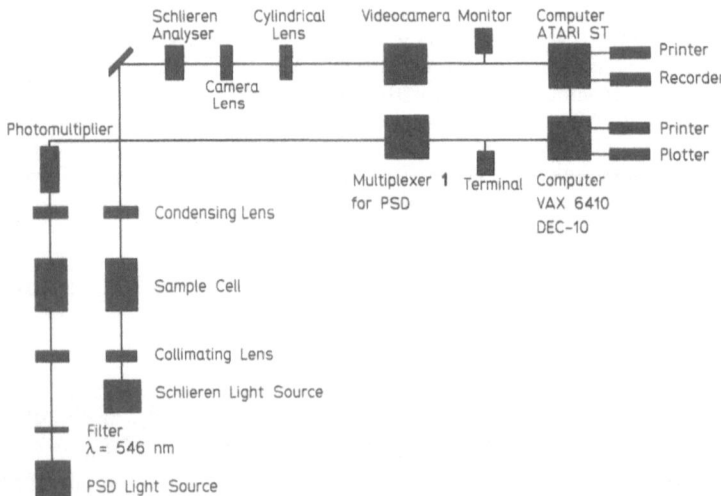

Fig. 2. Optical pathways of the second AUC system with video Schlieren optics (single method) and multiplexer system for PSD

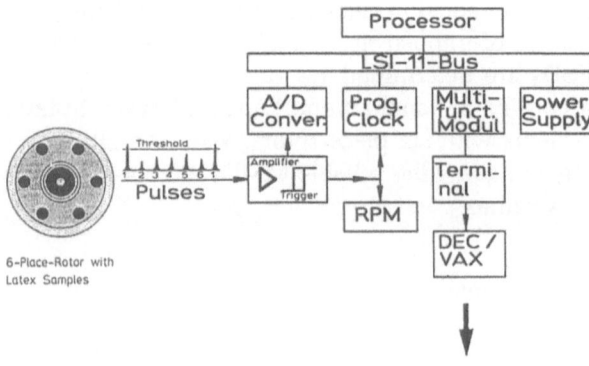

Fig. 3. Block diagram of the multiplexer systems for PSD

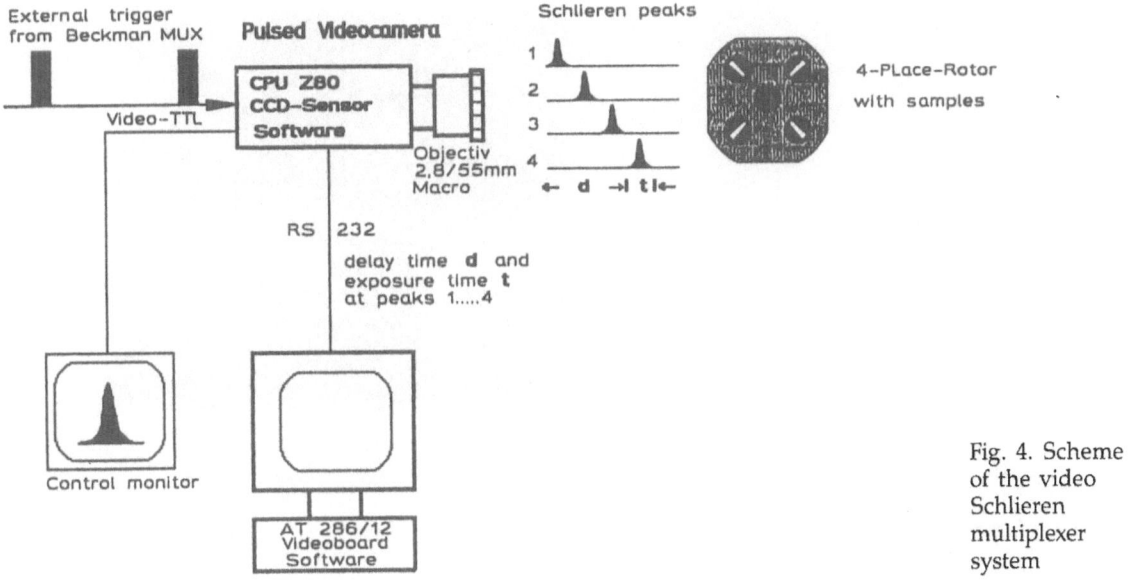

Fig. 4. Scheme
of the video
Schlieren
multiplexer
system

with the scanner slit position at the center of the cell. The original Beckman multiplexer triggers the video camera as well as the homemade PSD multiplexer system.

In principle, the second machine has the same optical pathways (Fig. 2). There are, however, only a simple video system for the Schlieren optics, no scanner, and a multiplexer only for the PSD analysis.

The multiplexer systems are shown in more detail in Fig. 3. Triggering is done either by an empty cell or by the coded rotor support ring, hence a six place rotor facilitates either five or six simultaneous measurements. A programmable clock (2 MHz) synchronizes the data collection with the peak maxima, a multifunctional board organizes the transfer of the A/D converted intensities to the central VAX computer and to a terminal, and the data evaluation is done on the VAX computer with custom software.

The function of the pulsed video camera (PCO, Kehlheim, FRG) is described in Fig. 4. Any continuous light source can be used. The supplied software synchronizes the exposure time exactly at the Schlieren peak; either sector in the double sector centerpieces can be selected, thereby eliminating background light via subtraction of the blank. The high sensitivity of the sensors facilitates very short exposure times; the time in which a cell is seen by the camera is in the range between 10 and 100 µs.

Very often, the Schlieren peaks are too small at the small concentrations required for reliable extrapolation to infinite dilution; therefore, we tried to change to the UV optics. This could only be done with a host of improvements: old electronic parts had to be replaced by selected transistors and diods, 1% tolerance metal film resistors, high grade capacitors, and ultra-low noise precision operational amplifiers; likewise, the photomultiplier

tube was replaced by a new one with high gain for UV radiation, and the new components had to be calibrated. Finally, new, coated aluminum mirrors with high UV reflectance were installed, thereby extending the range of wavelengths down to 210 nm; this is necessary because poly(meth)acrylates have to be measured at 230 nm or even below.

Results

Figure 5 gives, as examples, some PSD measurements that demonstrate the recording of small admixtures of different size, the high resolution, and a good reproducibility. Figure 6 shows a hardcopy of the monitor display of a Schlieren peak. Depending on the quality (i.e., contrast) of the recording the peak can be digitized automatically with custom computer software, or this must be done manually (with a computer mouse or by using a graphic tablet).

The data processing is depicted in Figs. 7 and 8 for PMMA in acetone as an example. After calculating the concentration gradients, one gets the $G'(S)$-function and, finally, the MMD (Fig. 9). This, however, is only an approximate one, because it was obtained from a finite concentration without extrapolation to infinite dilution, and is much too low.

Figure 10 shows that, at 230 nm, quite convincing results can be obtained on PMMA with the UV

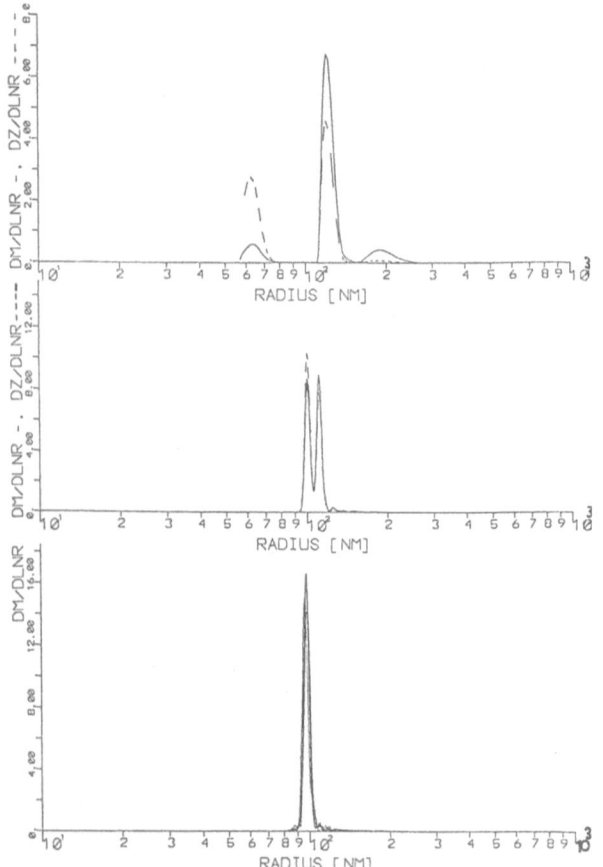

Fig. 5. PSDs of poly(butylacrylate) latexes showing (from top to bottom) the detection of small admixtures, the high resolution, and the good reproducibility (superposition of five runs). Solid lines show the mass distribution $dm/d\ln r$ and broken lines are the number distribution $dz/d\ln r$ of the particle radius r (where m and z are normalized masses and numbers, respectively)

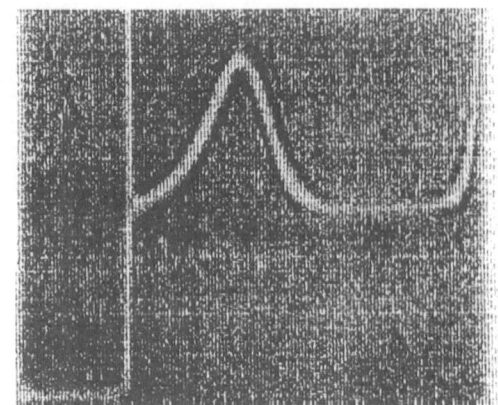

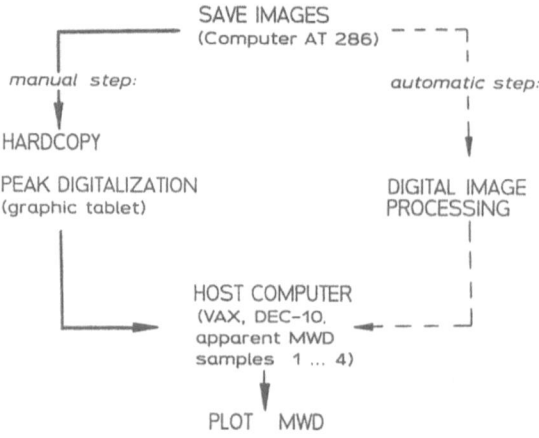

Fig. 6. Hardcopy of the monitor display of a Schlieren peak

optics. The now much smaller concentration yields almost the correct molecular mass without extrapolation (Fig. 11).

Outlook and conclusion

Further improvement of accuracy and speed will be achieved by installing collimating optics for the UV system, a laser for the Schlieren method, and an AT 486 personal computer with an i860 transputer board for the digital image processing. In favorable cases, a complete MMD will be obtained in less than 1 day, whereas with the conventional

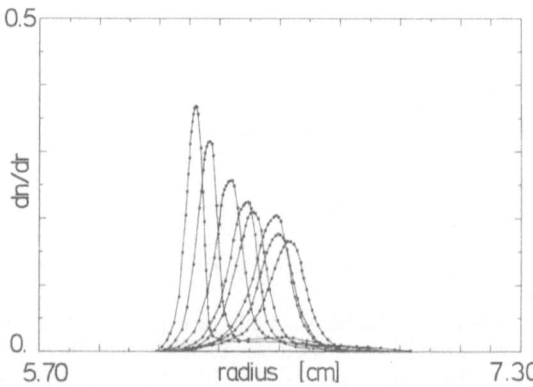

Fig. 7. Concentration gradients of poly(methylmethacrylate) in acetone measured at various times with a concentration c_0 of 2.0 mg/cm³

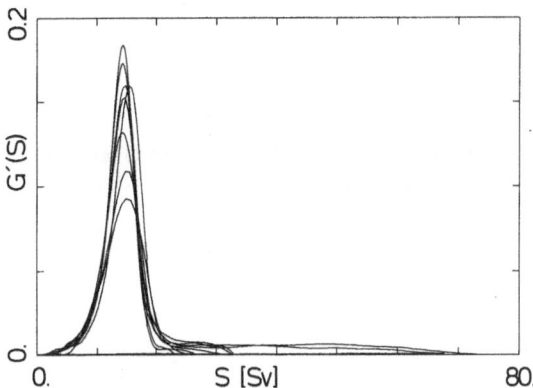

Fig. 8. Normalized apparent distributions G' as function of the sedimentation coefficient S at various times, calculated from Fig. 7

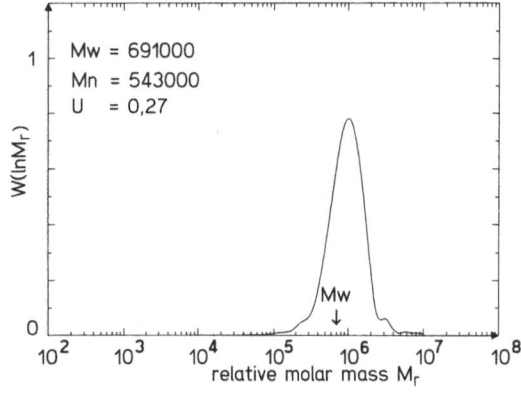

Fig. 11. Apparent MMD of PMMA, calculated from Fig. 10

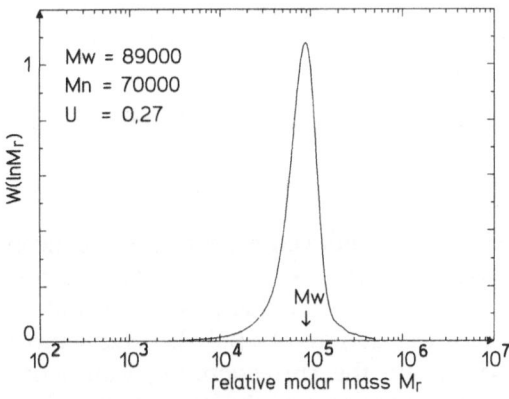

Fig. 9. Apparent MMD of PMMA at c_0 = 2.0 mg/cm³, calculated from Fig. 8

method without multiplexer and digital image processing [1] about 1 week was required. Thus, the multiplexer systems can reduce measuring time, thereby decreasing cost and expanding the range of applications in scientific and industrial research.

References

1. Stickler M (1984) Angew Makromol Chem 123/124:85—117
2. Scholtan W, Lange H (1972) Koll-Z Z Polym 250:782—796

Fig. 10. Concentration traces for PMMA (same sample as in Figs. 7—9) in acetonitril at various times, taken with the UV scanner (c_0 = 0.81 mg/cm³)

Authors' address:

Dr. D. Panke
Röhm GmbH, Abt. FE-S
Postf. 4242
W-6100 Darmstadt, FRG

Progress in Colloid & Polymer Science   Progr Colloid Polym Sci 86:62—69 (1991)

A new procedure for the determination of the molar mass distribution of polymers in solution from sedimentation equilibrium

M. D. Lechner and W. Mächtle*)

Physikalische Chemie, Universität Osnabrück, Osnabrück, FRG
*) Polymerphysik, Kunststofflaboratorium, BASF AG, Ludwigshafen, FRG

Abstract: A new method for the evaluation of multimodal molar mass distributions $W(M)$ from sedimentation equilibrium measurements of polymers in solution is proposed. The method is based on the determination of the experimental measured reduced concentration profile (interference optic) or reduced concentration gradient profile (Schlieren optic) by extrapolation to zero concentration, and subsequent direct calculation of the molar mass distribution by nonlinear regression of the integral equation. The procedure is tested with different polystyrenes and sodium polystyrene sulfonate.

Key words: Ultracentrifuge; sedimentation equilibrium; molar mass distribution; polymer solutions

Introduction

Besides scattering methods, osmotic pressure, viscosity and gel permeation chromatography (GPC), analytical ultracentrifugation (AUC) is a powerful tool for the determination of molar mass means and molar mass distributions. Compared with GPC the AUC is an absolute method and allows the estimation of a large range of molar masses by easy executable variation of the rotor speed. The object of the following presentation is the development of a new procedure for the determination of a fitted multimodal model molar mass distribution (Wesslau-, Schulz-Flory- and Poisson-type) from sedimentation equilibrium measurements. The method is checked with the polymer standards polystyrene and with the polyelectrolyte sodium polystyrene sulfonate.

Procedure

The determination of the molar mass distribution $W(M)$ and molar mass averages M_n, M_w, M_z, M_{z+1} requires the connection of the experimental measured polydisperse reduced concentration profile $C(X)/C_0$ with the molar mass distribution function $W(M)$ and the reduced monodisperse concentration profile $C_i(X)/C_{0,i}$. $C(X)$ and C_0 are the total solute concentration at a radial position X and the total initial concentrations of a polydisperse sample; $C_i(X)$ and $C_{0,i}$ are the corresponding values of a monodisperse sample or of component i in a polydisperse sample [1—5].

For polydisperse polymers in solution the following relations hold for the i-th component of the polymer [2]:

$$R\,dR(1 - v_2\rho_1)\omega^2/(RT)$$
$$= d\ln C_i/M_i + 2\sum_{k=1}^{q} A_{2ik}\,dC_k + \dots ;$$
$$(i = 1, 2, \dots, q) . \qquad (1)$$

Rearrangement in terms of the variable $X = (R^2 - R_m^2)/(R_b^2 - R_m^2)$ yields

$$\lambda\,dX = d\ln C_i/M_i + 2\sum_{k=1}^{q} A_{2ik}\,dC_k + \dots ;$$
$$(i = 1, 2, \dots, q) , \qquad (2)$$

with $\lambda = (1 - v_2\rho_1)\omega^2(R_b^2 - R_m^2)/(2\,RT)$.

R, R_m and R_b are the distances of the detected molecules, meniscus and bottom from center of rotation. The other constants and variables have their usual meaning. Division of Eq. (2) by $C_{0,i}$, the initial concentration of component i, yields for pseudoideal solutions (where $A_{2ik} = A_{3ik} = 0$) or for concentration profiles at infinite dilution $(C_i/C_{0,i})_{C_0 \to 0}$:

$$\lambda\, dX = d\ln U_i / M_i \; ;$$

$$(A_{2ik} = A_{3ik} = 0 \quad \text{or} \quad (C_i/C_{0,i})_{C_0 \to 0}) \; ;$$

$$(i = 1, 2, ..., q) \; , \tag{3}$$

where $U_i = C_i/C_{0,i}$ is the local concentration of solute i relative to its value in the initial solution; U_i is also called the monodisperse reduced concentration profile. With the condition for the conservation of mass

$$\int_0^1 C_i(X)\, dX = C_{0,i} \; ;$$

$$\int_0^1 U_i(X)\, dX = 1 \; ; \quad (i = 1, 2, ..., q) \tag{4}$$

integration of Eq. (3) gives

$$U_i(X) = \lambda M \exp(\lambda M X)/(\exp(\lambda M) - 1) \; ;$$

$$0 < X < 1 \; . \tag{5}$$

The total solute concentration C of a polydisperse sample at a radial position X may be obtained by summation over all solute components ($i = 1, 2, ..., q$)

$$C = \sum_{i=1}^{q} C_i = C_0 \sum_{i=1}^{q} W_i U_i \; ; \quad C_0 = \sum_{i=1}^{q} C_{0,i} \; , \tag{6}$$

with $W_i = C_{0,i}/C_0$ = weight fraction of solute i.

Combining Eqs. (5) and (6) and regarding W_i as a continuous function of M_i, the molar mass of the solute i, gives

$$U_w(X) = (C(X)/C_0)_{C_0 \to 0} = \int_0^\infty W(M) U(X, M)\, dM \tag{7}$$

$$U(X, M) = \lambda M \exp(\lambda M X)/(\exp(\lambda M) - 1) \; ;$$

$$0 < X < 1 \; . \tag{8}$$

Differentiation of Eq. (8) with respect to X yields

$$V_w(X) = ((dC(X)/dX)/C_0)_{C_0 \to 0}$$

$$= \int_0^\infty W(M) V(X, M)\, dM \tag{9}$$

$$V(X, M) = \lambda^2 M^2 \exp(\lambda M X)/(\exp(\lambda M) - 1) \; ;$$

$$0 < X < 1 \; . \tag{10}$$

For nonideal solutions, one has to perform a very precise extrapolation to infinite dilution or introduce correction terms. The problem of nonideal solutions may be avoided by measuring in pseudoideal solutions, but pseudoideal solutions are not available in all cases and, furthermore, this restricts the application. The theory of nonideal solutions with respect to sedimentation equilibrium is rather sophisticated [2], and equations for extrapolation of $C(X)/C_0$ to zero concentration are not given explicitly.

Therefore up to now we have been forced to use empirical relations for this extrapolation procedure. The extrapolation to infinite dilution was done for poor solvents by linear regression, whereas for good solvents we used quadratic regression.

There are two ways to get the molar mass distribution from the experimental measured concentration profiles. The first is the Laplace transformation of the Fredholm type integral equations Eqs. (9) and (7). The disadvantge of this mathematically strict procedure is that the operators are poorly conditioned, i.e., small changes of the experimental values exhibit large changes of the distribution function $W(M)$ and physically senseless negative values of $W(M)$ may occur [5, 6].

The other procedure is the direct calculation of the multimodal distribution function $W(M) = \sum g_i W(M)_i$ from $U_w(X)$ and $V_w(X)$ by nonlinear regression:

$$U_w(X) = (C(X)/C_0)_{C_0 \to 0} = \int_0^\infty W(M) U(X, M)\, dM$$

$$V_w(X) = ((dC(X)/dX)/C_0)_{C \to 0}$$

$$= \int_0^\infty W(M) V(X, M)\, dM \tag{11}$$

$$U_w(X) = f(X, k_1, ..., k_n)$$

$$V_w(X) = g(X, k_1, k_2, ..., k_n)$$

$$S = \sum (f(X_i, k_1, k_2, ..., k_n) - U_{w,i})^2$$

$$S = \sum (f(X_i, k_1, k_2, ..., k_n) - V_{w,i})^2$$

$k_1, k_2, ..., k_n = \vec{k} =$ constants of a molar mass distribution function i.e. $M_{n,i}, M_{w,i}, M_{z,i}, ..., g_i; C_i, B_i, K_i, ..., g_i$.

Suppose that $U_w(X)$ and $V_w(X)$ are functions of X and constants $\vec{k}$ which determine the multimodal mass distribution function $W(M)$. The nonlinear regression problem is then defined by minimizing S which is the squared sum of the differences between experimental points $U_{w,i}$ or $V_{w,i}$ and calculated points $f(X_i, \vec{k})$. The problem is solved in the usual way by calculating n partial derivatives of S with respect to $\vec{k}$ and setting them 0. The constants may be $M_{n,i}, M_{w,i}, M_{z,i}, ..., g_i$ or constants of a model molar mass distribution function e.g. $C_i, B_i, K_i, ..., g_i$ where g_i is the weighting factor for the construction of a multimodal distribution function $W(M)$ from a couple of monomodal distribution functions $W(M)_i$.

The calculation needs a model molar mass distribution $W(M)$. We propose the following distribution function $W(M)$:

$$W(M) = \sum g_i W(M)_i$$

with

$$W(M)_i = C1 M^K \exp(-BM^C) \tag{12}$$

$$C1 = C B^{K/C + 1/C} / \Gamma(K/C + 1/C)$$

$$M_n = \Gamma(K/C + 1/C) / (B^{1/C} \Gamma(K/C))$$

$$M_w = \Gamma(K/C + 2/C) / (B^{1/C} \Gamma(K/C + 1/C))$$

$$M_z = \Gamma(K/C + 3/C) / (B^{1/C} \Gamma(K/C + 2/C))$$

$$M_{z+1} = ...$$
$$\vdots$$

$\Gamma(X)$ = Gamma function

C = 0.5 Wesslau distribution, log-norm-distribution

C = 1   Schulz-Flory distribution
C = 2   Maxwell distribution, Poisson distribution.

This function covers nearly all types of distributions (Poisson, Schulz-Flory, Wesslau etc.) with different skewness.

This method of calculating $W(M)$ is, contrary to the inverse problem, well conditioned and very stable. The calculation can be performed mathematically by the multidimensional Newton method or by the multidimensional Simplex method. We preferred to use the Simplex method, because it needs no derivatives of the integral equation. The Newton method normally converges faster, but it is sometimes problematic, especially if the experimental values are noisy.

Experimental

The following experimental measurements demonstrate that the proposed procedure gives reasonable results. The experiments were performed at BASF (Ludwigshafen, FRG) during a guest stay. Experimental details of the procedure are given elsewhere [7—9]. The calculations were done with a personal computer with an 80286 processor and math coprocessor; the programs for calculation are available on request.

Figure 1 is a protocol of the investigation of sodium polystyrene sulfonate NaPSS 400000 (pressure chemical company) in 0.5 M NaCl solution. The data supplied by the manufacturer are $M_p = 400000$ g/mol; $M_w/M_n = 1.10$. The heading of the protocol contains constants and different values such as total fringe number J_0, fringe number at meniscus J_m, fringe number difference between bottom and meniscus dJ, and the appropriate concentrations, $M_{w,app}, M_{z,app}, M_w$, and A_2. The procedure of the calculation of the molar mass distribution is documented in Fig. 1. The extrapolation of $C(X)/C_0$ to zero concentration has been done by quadratic regression. The results show reasonable agreement with the values given by pressure chemical company.

The values of $M_{w,app}, M_{z,app}, M_w$, and A_2 at the top of the protocol are obtained from intercepts ($M_{w,app}$) and slopes ($M_{z,app}$) of $C(X)/C_0 = f(X)$ and subsequent extrapolation of $1/M_{w,app}$ to $C_0 \to 0$ [2, 8]. The reason for the relatively large inaccuracy of the $M_{z,app}$-values is that the slopes of the plot $C(X)/C_0 = f(X)$ can only be determined with a relatively large error. These calculations have nothing to do with the calculations of molar mass distribution by nonlinear regression, but it is sometimes useful to compare M_w and M_z obtained by both methods.

Figure 2 demonstrates measurements and calculations of the broadly distributed polystyrene NBS 706 (national bureau of standards; $M_n = 136000$ g/mol; $M_w = 258000$ to 288000 g/mol; $M_z:M_w:M_n = 2.9:2.1:1$). As is seen, the results of the calculations are in good agreement with values given by NBS. Figure 3 shows that polymers with low molar mass and small molar mass distribution

ULTRACENTRIFUGE (EQUILIBRIUM; INTERFERENCE-OPTIC)

Substanz: NaPSS 400000   Laufzeit: 71 h   Mess-Datum: 13.5.91   Auftrag Nr.:
Loesungsmittel: 0.5 M NaCl   Rotor: 8 L   Auswerte-Datum: 13.5.91   Platten-Nr.: 4352
Bemerkungen.: AUZ 2

Drehzahl in 1/min: 2473   Dichte LM in g/ccm: 1.0173   Spez.Vol. PM in ccm/g: .626
dn/dC in ccm/g: .173   Wellenlaenge in cm: .0000633 Temperatur in K: 298
Vergr. der UZ-Optik MX: 2.1573 Laenge l in cm: 1.2   Faktor f. Konzentr.:   1
Profil-Proj. Vergr.:   10.039 Eichkonst. in 1/cm: .00234

Zei	c/(g/l)	Rm/cm	Rb/cm	Jo	Jm	dJ	Cb-Cm	Cm	Mw,app	Mz,app	xxxxxx
□	.507	6.87	7.074	1.662	1.208	.94	.286	.368	404300	-18700	
+	1.05	6.851	7.069	3.443	2.757	1.576	.48	.84	307200	763000	
*	1.98	6.83	7.046	6.493	5.46	2.184	.666	1.664	228400	79600	
x	2.99	6.86	7.075	9.806	8.28	3.117	.95	2.524	216100	2000	
△	3.99	6.855	7.068	13.085	11.622	3.072	.936	3.543	161400	224200	
◇	5.02	6.847	7.058	16.463	14.801	3.356	1.023	4.513	141900	59500	
▽	5.98	6.842	7.055	19.612	17.691	3.761	1.147	5.394	132100	8700	

Mw und A2 aus 1/Mw, app = f((Cm + Cb)/2): Mw = 457000 g/mol ;   A2*10^4 = 4.68 cm^3 mol/g^2
Mw und A2 aus 1/Mw, app = f(Co/2):   Mw = 452000 g/mol ;   A2*10^4 = 4.66 cm^3 mol/g^2

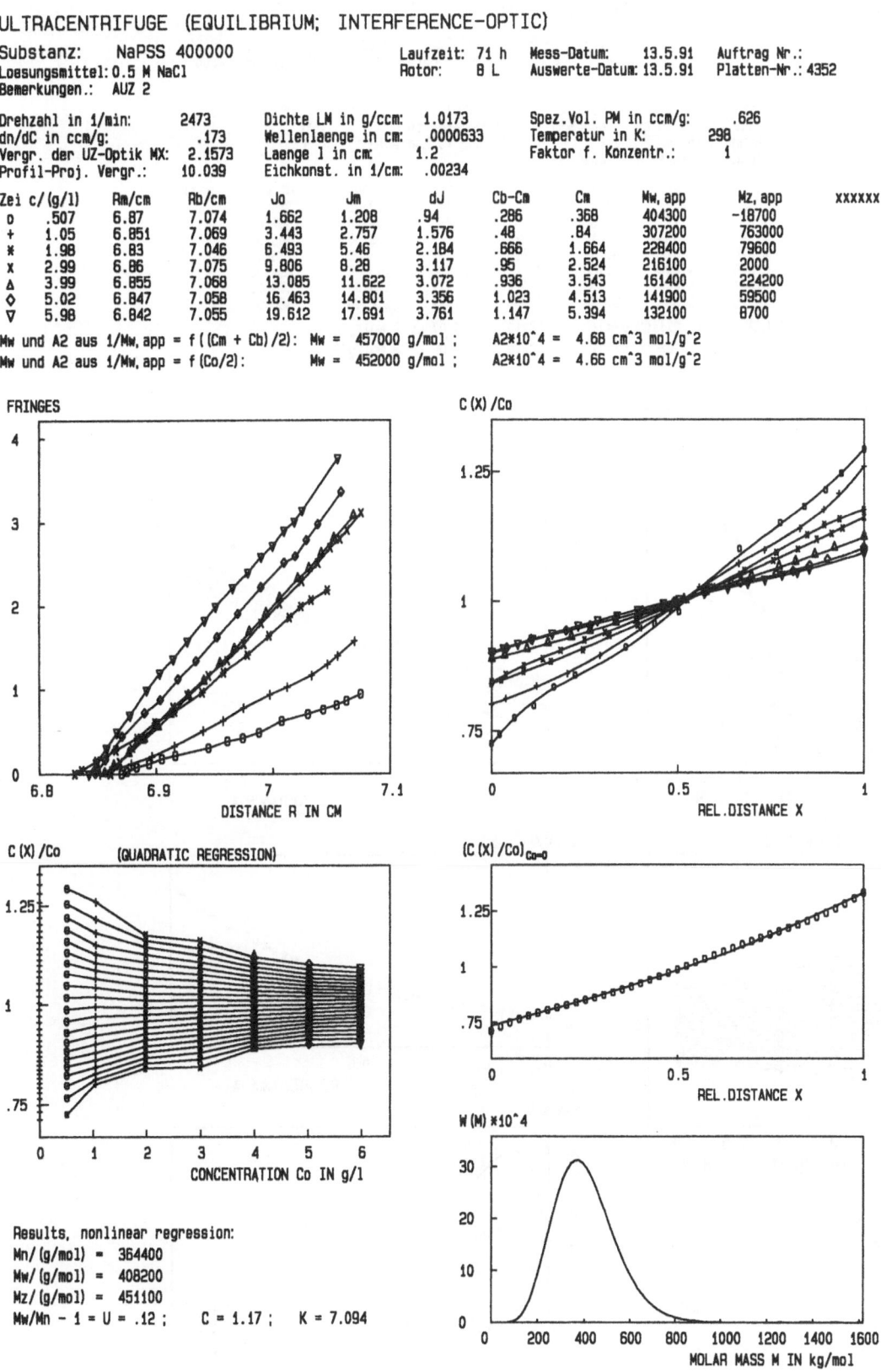

Results, nonlinear regression:
Mn/(g/mol) = 364400
Mw/(g/mol) = 408200
Mz/(g/mol) = 451100
Mw/Mn - 1 = U = .12 ;   C = 1.17 ;   K = 7.094

Fig. 1. Protocol of sodium polystyrene sulfonate NaPSS 400000 in 0.5 M NaCl aqueous solution; interference optic

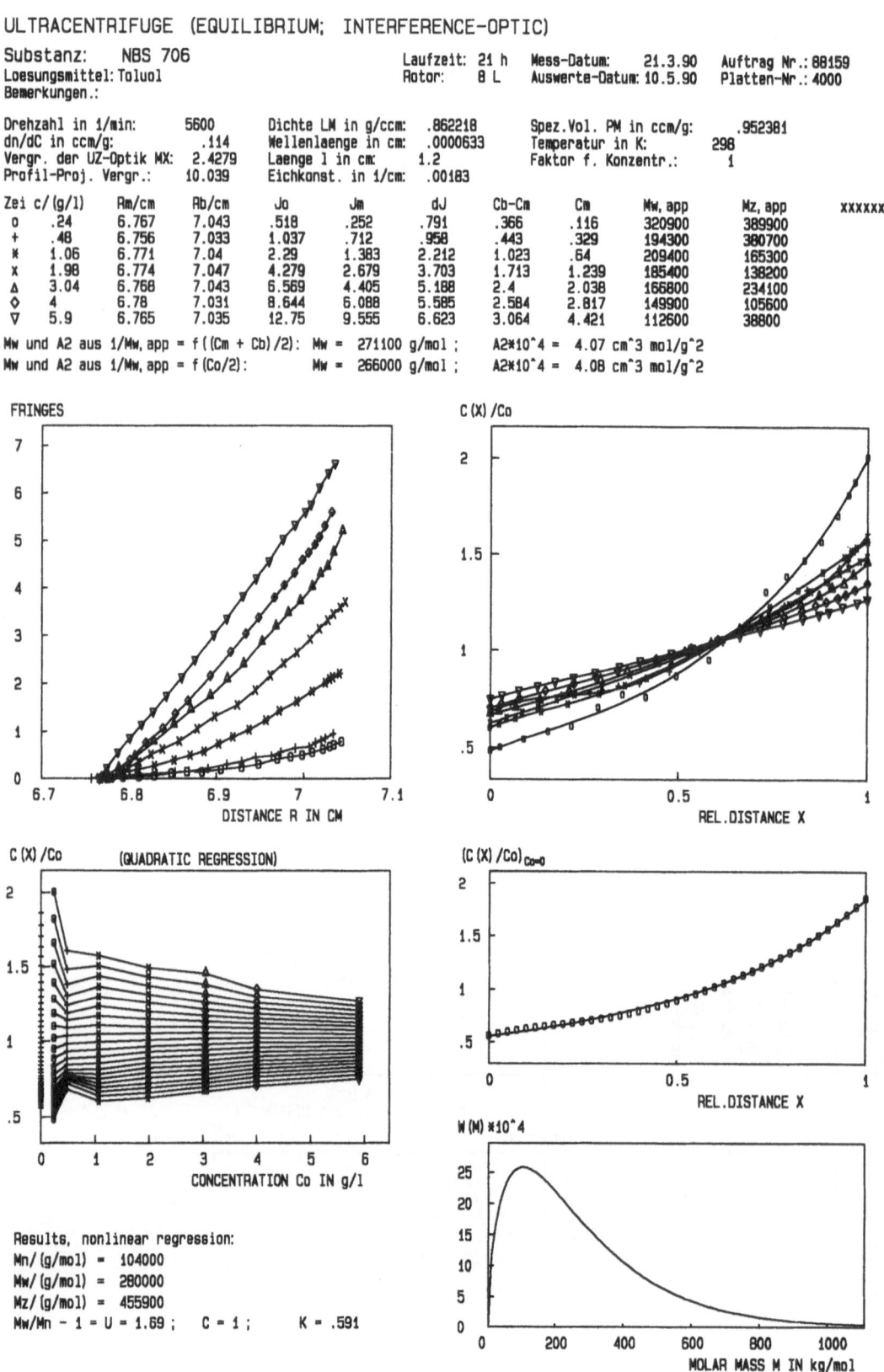

ULTRACENTRIFUGE (EQUILIBRIUM; INTERFERENCE-OPTIC)

Substanz: NBS 706   Laufzeit: 21 h   Mess-Datum: 21.3.90   Auftrag Nr.: 88159
Loesungsmittel: Toluol   Rotor: 8 L   Auswerte-Datum: 10.5.90   Platten-Nr.: 4000
Bemerkungen.:

Drehzahl in 1/min: 5600   Dichte LM in g/ccm: .862218   Spez.Vol. PM in ccm/g: .952381
dn/dC in ccm/g: .114   Wellenlaenge in cm: .0000633 Temperatur in K: 298
Vergr. der UZ-Optik MX: 2.4279   Laenge l in cm: 1.2   Faktor f. Konzentr.:   1
Profil-Proj. Vergr.:   10.039   Eichkonst. in 1/cm: .00183

Zei	c/(g/l)	Rm/cm	Rb/cm	Jo	Jm	dJ	Cb-Cm	Cm	Mw,app	Mz,app	
o	.24	6.767	7.043	.518	.252	.791	.366	.116	320900	389900	xxxxxx
+	.48	6.756	7.033	1.037	.712	.958	.443	.329	194300	380700	
✳	1.06	6.771	7.04	2.29	1.383	2.212	1.023	.64	209400	165300	
x	1.98	6.774	7.047	4.279	2.679	3.703	1.713	1.239	185400	138200	
△	3.04	6.768	7.043	6.569	4.405	5.188	2.4	2.038	166800	234100	
◇	4	6.78	7.031	8.644	6.088	5.585	2.584	2.817	149900	105600	
▽	5.9	6.765	7.035	12.75	9.555	6.623	3.064	4.421	112600	38800	

Mw und A2 aus 1/Mw,app = f((Cm + Cb)/2): Mw = 271100 g/mol ;   A2*10^4 = 4.07 cm^3 mol/g^2
Mw und A2 aus 1/Mw,app = f(Co/2):   Mw = 266000 g/mol ;   A2*10^4 = 4.08 cm^3 mol/g^2

FRINGES

C(X)/Co

C(X)/Co   (QUADRATIC REGRESSION)

(C(X)/Co)$_{Co=0}$

Results, nonlinear regression:
Mn/(g/mol) = 104000
Mw/(g/mol) = 280000
Mz/(g/mol) = 455900
Mw/Mn - 1 = U = 1.69 ;   C = 1 ;   K = .591

W(M) *10^4

Fig. 2. Protocol of polystyrene NBS 706 in toluene; interference optic

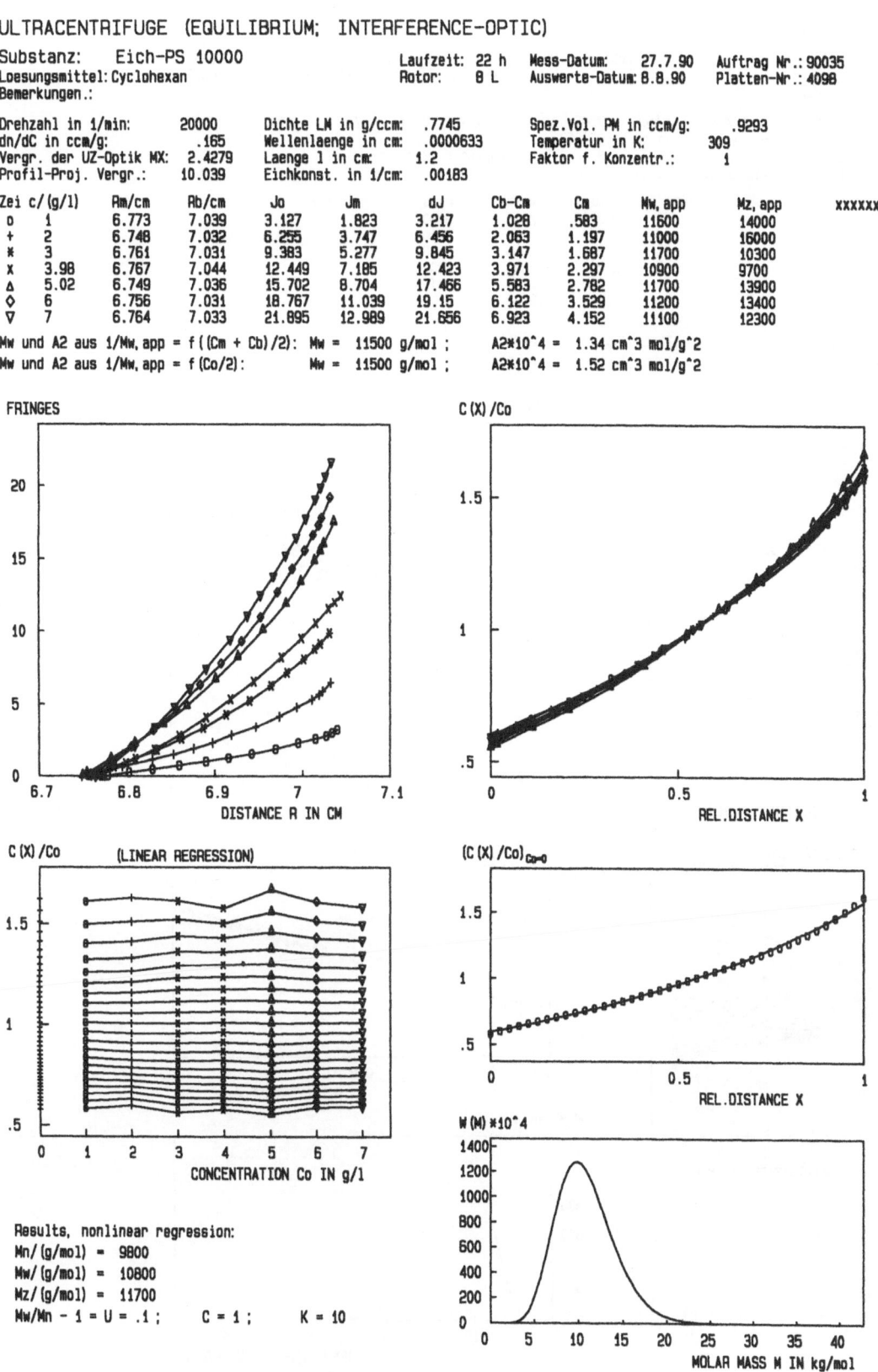

```
ULTRACENTRIFUGE (EQUILIBRIUM; INTERFERENCE-OPTIC)

Substanz:      Eich-PS 10000          Laufzeit: 22 h    Mess-Datum:      27.7.90    Auftrag Nr.: 90035
Loesungsmittel: Cyclohexan            Rotor:     B L    Auswerte-Datum: 8.8.90     Platten-Nr.: 4098
Bemerkungen.:

Drehzahl in 1/min:        20000    Dichte LM in g/ccm:  .7745    Spez.Vol. PM in ccm/g:   .9293
dn/dC in ccm/g:             .165   Wellenlaenge in cm:  .0000633 Temperatur in K:      309
Vergr. der UZ-Optik MX:    2.4279  Laenge l in cm:       1.2     Faktor f. Konzentr.:      1
Profil-Proj. Vergr.:      10.039   Eichkonst. in 1/cm:  .00183

Zei c/(g/l)  Rm/cm   Rb/cm    Jo       Jm       dJ      Cb-Cm    Cm      Mw,app    Mz,app     xxxxxx
 □   1       6.773   7.039    3.127    1.823    3.217   1.028    .583    11600     14000
 +   2       6.748   7.032    6.255    3.747    6.456   2.063    1.197   11000     16000
 *   3       6.761   7.031    9.383    5.277    9.845   3.147    1.687   11700     10300
 x   3.98    6.767   7.044   12.449    7.185   12.423   3.971    2.297   10900     9700
 △   5.02    6.749   7.036   15.702    8.704   17.466   5.583    2.782   11700     13900
 ◇   6       6.756   7.031   18.767   11.039   19.15    6.122    3.529   11200     13400
 ▽   7       6.764   7.033   21.895   12.989   21.656   6.923    4.152   11100     12300

Mw und A2 aus 1/Mw,app = f((Cm + Cb)/2):  Mw = 11500 g/mol ;   A2*10^4 = 1.34 cm^3 mol/g^2
Mw und A2 aus 1/Mw,app = f (Co/2):        Mw = 11500 g/mol ;   A2*10^4 = 1.52 cm^3 mol/g^2
```

FRINGES

DISTANCE R IN CM

C (X) /Co

REL.DISTANCE X

C (X) /Co   (LINEAR REGRESSION)

CONCENTRATION Co IN g/l

(C (X) /Co)$_{Co=0}$

REL.DISTANCE X

Results, nonlinear regression:
Mn/(g/mol) = 9800
Mw/(g/mol) = 10800
Mz/(g/mol) = 11700
Mw/Mn - 1 = U = .1 ;   C = 1 ;   K = 10

W(M) *10^4

MOLAR MASS M IN kg/mol

Fig. 3. Protocol of polystyrene PCC 8b in cyclohexane; interference optic

ULTRACENTRIFUGE (EQUILIBRIUM; SCHLIEREN-OPTIC)

Substanz:	Eich-PS 10000	Laufzeit: 20 h	Mess-Datum:	26.7.90	Auftrag Nr.: 90035
Loesungsmittel: Cyclohexan		Rotor: 8 L	Auswerte-Datum: 12.9.90		Platten-Nr.: 4097S
Bemerkungen.: kuenst.Boden H2O					

Drehzahl in 1/min:	14000	Dichte LM in g/ccm:	.7647	Spez.Vol. PM in ccm/g:	.931
dn/dC in ccm/g:	.163	Wellenlaenge in cm:	.0000633	Temperatur in K:	309
Vergr. der UZ-Optik MX:	2.4279	Laenge l in cm:	1.2	Faktor f. Konzentr.:	1
Profil-Proj. Vergr.:	10.039	Eichkonst. in 1/cm:	.00183		

Zei	c/(g/l)	Rm/cm	Rb/cm	Cb-Cm	Cm	Cb	Mw,app	Mz,app	xxxxxx	xxxxxx
□	2	6.74	7.021	.606	1.722	2.328	12300	9000		
+	3	6.755	7.022	1.01	2.524	3.534	13000	7200		
✳	3.98	6.762	7.038	1.19	3.423	4.614	12500	6500		
×	5.02	6.748	7.028	1.557	4.303	5.861	12300	8400		
△	6	6.754	7.026	1.805	5.18	6.985	12700	10100		
◇	7	6.764	7.029	2.035	6.081	8.117	13000	10900		

Mw und A2 aus 1/Mw,app = f((Cm + Cb)/2): Mw = 12300 g/mol ;   A2*10^4 = -.01 cm^3 mol/g^2
Mw und A2 aus 1/Mw,app = f(Co/2):   Mw = 12300 g/mol ;   A2*10^4 = -.01 cm^3 mol/g^2

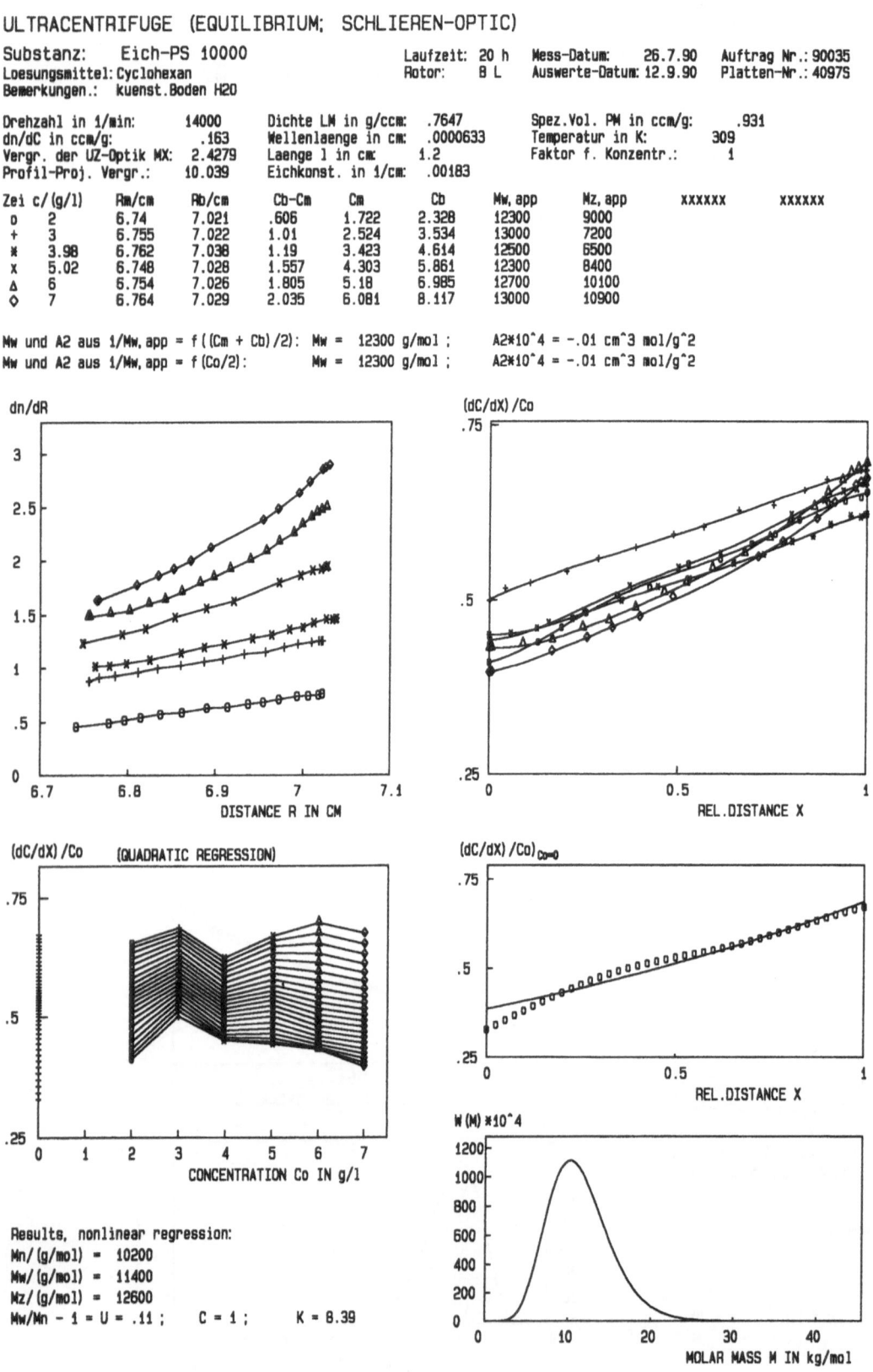

Results, nonlinear regression:
Mn/(g/mol) = 10200
Mw/(g/mol) = 11400
Mz/(g/mol) = 12600
Mw/Mn - 1 = U = .11 ;   C = 1 ;   K = 8.39

Fig. 4. Protocol of polystyrene PCC 8b in cyclohexane; Schlieren optic

can also be investigated. The obtained molar mass M_w = 10800 g/mol and the polydispersity M_w/M_n = 1.1 for polystyrene PCC 8b (pressure chemical company) are in good agreement with values given by the manufacturer (M_p = 10000 g/mol; M_w/M_n = 1.05).

Although the accuracy of a Schlieren optic is five to 10 times smaller than the interference optic, it is also possible to determine the molar mass distribution of polymers with Schlieren optics. Figure 4 shows measurements of polystyrene PCC 8b (pressure chemical company; M_p = 10000 g/mol; M_w/M_n = 1.05) in cyclohexane. The agreement with the values given by pressure chemical company is reasonable.

The study demonstrates, that equilibrium measurements with the AUC is a powerful tool for the determination of molar mass distributions of polymers. The results of the calculations may be improved by improving the precision of the experimental values. In this way, it might be possible to determine multimodal molar mass distribution, as is shown in the theoretical part of this paper, but up to now has not been experimentally verified. In the near future, we plan to analyze multimodal polymers.

Acknowledgement

Support of this work by BASF AG, Ludwigshafen and by the Fonds der Chemischen Industrie is gratefully acknowledged.

References

1. Fujita H (1962) Mathematical Theory of Sedimentation Analysis. Academic Press, New York
2. Fujita H (1975) Foundations of Ultracentrifugal Analysis. Wiley, New York
3. Scholte ThG (1968) J Polym Sci A2, 6:91, 111
4. Scholte ThG (1970) Eur Polym J 6:51
5. Greschner GS (1984) Eur Polym J 20:475
6. Wiff DR (1973) J Polym Sci, Polym Symp 43:219
7. Mächtle W, Klodwig U (1979) Makromol Chem 180:2507
8. Lechner MD, Mächtle W (1991) Makromol Chem 192:1183
9. Klodwig U, Mächtle W (1989) Coll Polym Sci 267:1117

Authors' address:

Prof. Dr. M. D. Lechner
Physikalische Chemie
Universität Osnabrück
Barbarstraße 7
4500 Osnabrück, FRG

Progress in Colloid & Polymer Science   Progr Colloid Polym Sci 86:70—75 (1991)

Determination of the degree of swelling and crosslinking of latex particles by analytical ultracentrifugation

H. G. Müller*), A. Schmidt, and D. Kranz

Bayer AG, Leverkusen, FRG
*) Bayer AG, ZF-TPP 5; E-41, Leverkusen, FRG

Abstract: Analytical ultracentrifugation has been used to characterize the solution properties of latices from emulsion polymerization. For that purpose, latices have to be taken over into a thermodynamically good solvent, and the distribution of sedimentation coefficients (s-values) has to be measured, consisting of high s-values due to swollen particles and low s-values from soluble macromolecules. From that, the portion of soluble macromolecules and swollen particles is available and from the s-value of the swollen particles their degree of swelling can be determined from fresh latex samples.

Key words: Latex particles; degree of swelling; distribution of sedimentation coefficients; analytical ultracentrifugation

1. Introduction

Many polymers, especially rubber polymers, are partially crosslinked. The degree of crosslinking is usually determined from polymer films. This procedure is also often applied to polymers which are made by emulsion polymerization. In this case, however, another method can be used which eliminates the influence of film formation on crosslinking. This method studies the crosslinking of latex particles which have just left the emulsion polymerization vessel. This may be done by adding an organic solvent to the latex, followed by preparative ultracentrifugation and the determination of the soluble part and the degree of swelling of the gel. The conditions of this type of ultracentrifugation are chosen more or less arbitrarily, which may influence the results considerably; furthermore, the degree of swelling thus obtained is only an average value.

Therefore, a more detailed analysis of the swelling of the latex particles is desirable. This can be achieved by the determination of the distribution of the sedimentation coefficients (s-distribution) of the latex particles in a thermodynamically good solvent by analytical ultracentrifugation.

This implies, however, that the s-distribution in the organic solvent is that of the unagglomerated single particles. Therefore, the latex particles have to be taken over into the organic solvent without agglomeration first.

2. Methods of transfer

Four methods of transferring the latex particles from the aqueous to the organic phase are possible.

1) Precipitation of latex into acetone or methanol, followed by instantaneously adding the precipitate to a good organic solvent, and removal of water, using zeolites.
2) The dropwise addition of latex to a boiling organic solvent and removal of water by azeotropic distillation.
3) The addition of organic solvent to the aqueous latex, if the organic solvent and water are miscible.
4) Freeze-drying of a latex to a powder and redispersion of the powder to an organic solvent 1).

The third method has the disadvantage that there is always a certain percentage of water in the organic solvent that reduces its thermodynamic quality and the swelling of the particles.

The second method may lead to an increase of crosslinking of the particles caused by the elevated temperature of the boiling azeotrope.

All three methods may lead to an (at least predominant) organic phase, in which the latex particles are individually suspended. This is proven by the fact that if a system of highly crosslinked particles is re-suspended in water, the latex has approximately the same particle size distribution as the original latex.

3. Method of measurement

The method of determining the *s*-distribution is the conventional determination of *s*-coefficients by analytical ultracentrifugation via interference optics [2, 3], except that we use wedge windows in order to have a five-hole multiplexer. Measurements are carried out at at least two speeds of the rotor, at low speed, ca. 2000 rpm to measure the fast sedimenting component (swollen particles), and at high speed ca. 40000 rpm for the soluble macromolecules.

4. Types of results

Depending on the degree of crosslinking, the behavior of the latex particles in the organic solvent is different (see Fig. 1).

In Fig. 1, four types of latex particles with identical diameter in water, but with different degrees of crosslinking (first column) have been compared. The second column illustrates what happens if these particles are transferred to a thermodynamically good solvent; the third shows the distribution of sedimentation coefficients in this solvent, as measured by analytical ultracentrifugation.

The particles of the latex in the first line are not at all crosslinked. Thus, they are totally soluble in the organic solvent and result in a distribution of low sedimentation coefficients, typical for soluble individual macromolecules. Totally crosslinked particles in line four show a contrary behaviour. There is no solubilization, not even swelling in the organic solvent, and the *s*-distribution (third column) shows only fast sedimentation coefficients. The

partly crosslinked latices (lines 2 and 3) show an intermediary behaviour: The soluble macromolecule still have low *s*-values, but these values increase with a rise in crosslinking, due to an increase in their molecular weight.

This rise of crosslinking also causes a decrease in the portion of soluble polymer. Inversely, the portion of the swollen particles increases. The rise of crosslinking leads simultaneously to an increase of the *s*-value, due to a reduction of swelling.

5. Evaluation of results

From a typical distribution of sedimentation coefficients for the crosslinked latex in Fig. 2, the following information is obtained:

1) The portion of the dissolved macromolecules and of the crosslinked particles in the sample: as a rule, the higher the portion of soluble macromolecules, the lower the degree of crosslinking of the latex particles.
2) The *s*-distribution of the dissolved macromolecules: the higher the *s*-values, the higher the molecular weight and/or the degree of branching.
3) From the *s*-values of the fast sedimenting particles the degree of swelling of the particles can be calculated. To do this, the absolute particle diameter of the unswollen latex has to be determined first, e.g., by analytical ultracentrifugation [4—7]. From that diameter the *s*-value of the latex in the organic solvent can be found by the following equation:

$$s = \frac{d_{TK}^2(\rho_{TK} - \rho_0)}{18 \cdot \eta} ; \qquad (1)$$

s = sedimentation coefficient;
d_{TK} = diameter of the compact, unswollen particles;
ρ_{TK} = density of the compact, unswollen particles;
ρ_0 = density of the dispersion medium;
η = viscosity of the diluted dispersion,

assuming that the particles are totally crosslinked and no swelling takes place.

In the case of swelling, if there is no soluble portion of the latex particle, the degree of swelling Q is

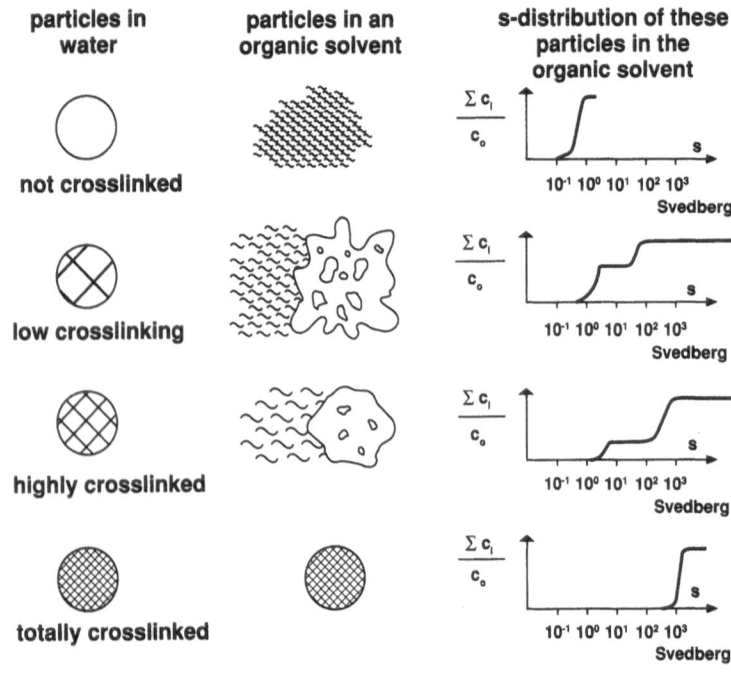

Fig. 1. Distribution of the sedimentation coefficient of latices with different degrees of crosslinking in a good solvent

$$Q = \left(\frac{s_{\text{unswollen}}}{s_{\text{swollen}}} \right)^3 ; \tag{2}$$

$s_{\text{unswollen}}$ = sedimentation coefficient of the unswollen, compact particles;
s_{swollen} = sedimentation coefficient of the swollen particles.

Q is, as usual, the volume ratio of the swollen and the unswollen, compact particles [8].

This is because the force of sedimentation is independent on swelling, whereas the force of friction follows Stoke's law.

$$f_f = 6 \cdot \pi \cdot R_{sw} \cdot \eta \cdot \frac{dx}{dt} ; \tag{3}$$

f_f = force of friction;
R_{sw} = radius of the swollen particles;
η = viscosity of the diluted dispersion;
$\frac{dx}{dt}$ = velocity of the particles.

The situation changes if a portion of the latex particle is soluble. Then the force of sedimentation also reduces, because the mass of the particles is reduced by a factor f:

$$m_r = m \cdot f ; \tag{4}$$

m_r = mass of the particle, reduced by the soluble part;
m = mass of the particle consisting of the soluble and insoluble part;
f = factor.

Applying the interference method of ultracentrifugation, f can easily be figured by

$$f = \frac{\sum i - \sum i_{\text{low}}}{\sum i} ; \tag{5}$$

where

$\sum i$ = total number of interference fringes, and
$\sum i_{\text{low}}$ = number of the interference fringes of the low sedimentation coefficients.

Assuming that the hydrodynamic diameter of the particle is not affected by the process of solubilization (this is an approximation of course), it follows that

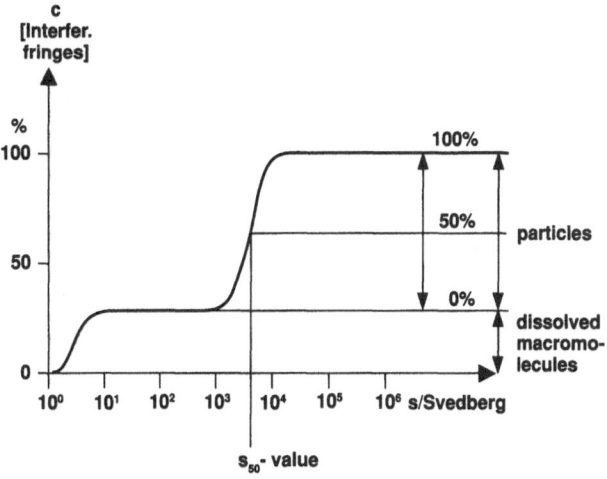

Fig. 2. Integral distribution of the sedimentation coefficient for a crosslinked latex in an organic solvent

$$Q = \left(\frac{1}{s_{\text{swollen}}} \cdot f \cdot \frac{\rho_{TK} - \rho_0}{\eta} \cdot \frac{1}{18} \cdot d_{TK}^2 \right)^3 ; \quad (6)$$

ρ_{TK} = density of the compact, unswollen particles;

d_{TK} = diameter of the compact, unswollen particles.

The Q values may provide a basis to calculate the effective network chain density following [9, 10], or the average degree of polymerization P_c of the molecule chains between two adjacent sites of crosslinking [11—13].

Another approximation comes from the sedimentation coefficient, determined in this work at a polymer concentration of 5 g/l, which has been inserted into Eq. (6) instead of the sedimentation constant. The sedimentation constant is calculated from the sedimentation coefficient by extrapolating it to zero concentration. In the following this value of the degree of swelling is abbreviated as Q^* (cp. Fig. 3); it is called here nominal Q.

As a rule, Q^* is bigger than Q, due to the considerable concentration dependence of the sedimentation coefficient.

6. Examples

In the following, two examples of the application of this method are given.

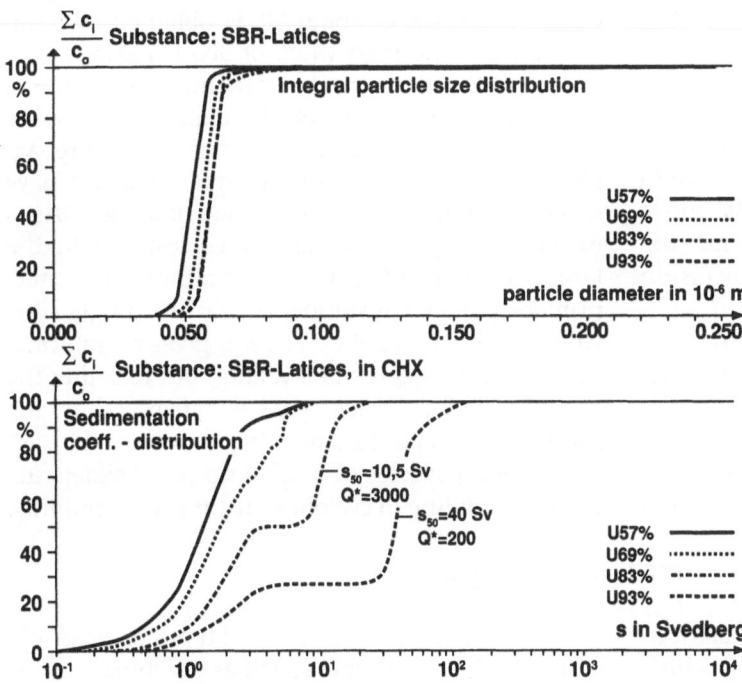

Fig. 3a, b. Particle size distribution and s-distribution of latices of the same polymerization process with increasing conversion U

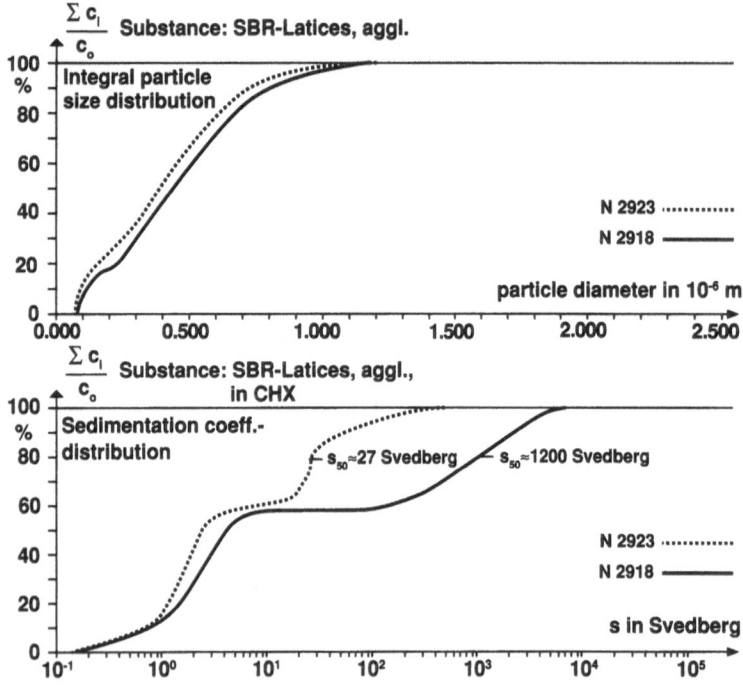

Fig. 4a, b. Particle size distribution and s-distribution of two agglomerated latices of different stability

The first example refers to styrene — butadiene latices from a batch process.

Samples taken with increasing degree of conversion are analysed as to their initial particle size distributions, thereafter transferred from water to cyclohexane, and reexamined regarding their s-distribution. Results are given in Fig. 3a and b.

Figure 3a shows that after a conversion of 70% has been reached, the particle size distribution of the latex practically no longer changes.

But the s-distribution of the samples in cyclohexane (Fig. 3b) is different. With increasing conversion first the monomodal s-values increase, due to an increase in molecular weight, then at 83% the s-distribution becomes bimodal, the high s-values being due to the swollen particles. At 93% conversion this portion is about 70%.

The degree of swelling, figured from Eq. (6) drops from 3000 (83% conversion) to 200 (93% conversion). The high values for the nominal Q are due to the fact that the concentration dependence of s has been neglected. This is an impressive example of the increase of crosslinking during an emulsion polymerization.

The second example shows that this method can detect small differences in latex stability.

Figure 4a shows the broad particle size distributions of two pressure agglomerated latices which are practically identical. After the continuous phase of these latices has been changed from water to cyclohexane, their s-distribution is as shown in Fig. 4b. Both distributions are bimodal but the s_{50}-values of the swollen particles differ greatly. As a result, the degree of swelling (nominal Q) of sample N 2918 turns out to be 300, an usual value; in contrast, a value of about 10^8 is obtained for sample N 2923. The high value of 300 is again due to neglect of the concentration dependence of s, but at first glance, the value of 10^8 makes no physical sense. The explanation is that the pressure agglomerated particles of sample N 2923 have dispersed into its primary small particles. As a result, the s_{50}-value does not correspond to the large agglomerated particles, but rather to the small primary particles resulting in this exeedingly high value for the nominal Q. Thus, it gives no information about the degree of swelling, but tells that the agglomerated particles have disintegrated in the presence of cyclohexane. In contrast, the agglomerated particles of N 2918 do not disintegrate, even on standing in cyclohexane for several months.

7. Conclusion

This method is applicable to all types of crosslinked latices, homopolymers such as polybutadiene or polychloroprene, copolymers such as poly (buta-diene-co-styrene) and graftlatices. Table 1 shows the advantages offered by this method.

Table 1. Advantages of the method

— Analysis of the degree of crosslinking using the fresh latex sample;

— Errors by complicated procedures of sample preparation are excluded;

— Comprehensive analysis of particles:
  — soluble portion/disperse portion
  — degree and distribution of crosslinking

References

1. Laun HM (1984) Die Angew Makromol Chem 123/124:335
2. Model E, Beckman, Munich
3. Chervenka CH (1969) A Manual of methods. Spinco Division of Beckman Instr., Palo Alto, California, p 23
4. Müller HG (1989) Colloid Polym Sci 267:1113
5. Scholtan W, Lange H (1972) Kolloid-Z u. Z Polymere 250:782
6. Mächtle W (1988) Die Angew Makromol Chem 162:35
7. Mächtle W (1984) Makromol Chem 185:1025
8. Lange H (1986) Colloid Polym Sci 264:488
9. Lorente MA, Mark JE (1980) Macromolecules 13:681
10. Mark JE (1981) Pure Appl Chem 53:1495
11. Flory PJ, Rehner Jr J (1943) J Chem Phys 11:521
12. Flory PJ (1950) J Chem Phys 18:108
13. Rehage G (1977) Berichte der Bunsengesellschaft Bd 81, 10:969

Authors' address:

Dr. H. G. Müller
Bayer AG; ZF-TPP 5, E-41
5090 Leverkusen, FRG

Determination of coalescence stability of emulsions by analytical ultracentrifugation under separation of dispersed phase*)

K. Strenge and A. Seifert[1])

Central Institute of Physical Chemistry, Berlin-Adlershof, FRG
[1]) Present address: Central Institute of Nutrition, Potsdam-Rehbrücke, FRG

Abstract: A centrifugal method for determining the coalescence stability of emulsions by analytical ultracentrifugation under separation of dispersed phase is described. The non-kinetically oriented method involves demulsification up to constant heights (final values) of all the layers. It can be performed in various variants including stepwise centrifugation that provides several final values during one run. On this basis, coalescence pressures are calculated as a measure of coalescence stability. The results obtained are in agreement with the expected behavior of the emulsions. In the case of SDS stabilized n-decane-in-water emulsions without electrolyte four layers are obtained that exist only during centrifugation under a sufficient centrifugal force, but not under gravity. A proof is given of a pressure barrier at the transition of the non-transparent to the transparent emulsion. The possible influence of the particle size on the coalescence pressures is discussed.

Key words: Emulsions; coalescence stability; demulsification; analytical ultracentrifugation; pressure barrier

Introduction

Centrifugal methods for determining the coalescence stability are widely used today. In most of these methods measures of stability are derived from the demulsification process under a constant centrifugal force. Kinetic and non-kinetic methods can be distinguished. In the former method stability indices are derived *during* the demulsification process. In the non-kinetic method applied in our investigations, indices are obtained *after finishing* the non-complete demulsification of the emulsion sample in the ultracentrifuge cell [1, 2]. This method will be called in the following "final" or "equilibrium" value method.

In 1943, Merill [3] reported on the determination of the mechanical stability of O/W emulsions with a laboratory centrifuge. The method involves measuring the rate of separation of dispersed phase, and the reciprocal of the initial rate of separation at a constant rotor speed has been taken as a quantitative index of the mechanical stability of the emulsions. In 1962, the first investigations with Beckman Model E Analytical Ultracentrifuges were published by Rehfeld [4], Garrett [5], Vold and Groot [6]. They used various indices to evaluate the coalescence stability, e.g., the volume fraction of remaining emulsion after various times of centrifugation. During the following years up to the present many papers on the application of these kinetically oriented methods with some further measures of stability have been published. Molecular mechanisms of coalescence and demulsification under centrifugal fields have been investigated and discussed, e.g., in [7, 13, 15]. The extensive in-

---
*) This paper was presented at the "VII. Symposium on Analytical Ultracentrifugation", Duisburg, FRG, February 28 — March 1, 1991.

vestigations of Vold and his co-workers Groot, Mittal, Hahn, Maletic, and Acevedo should be pointed out. These works have been excellently reviewed by Vold, Mittal and Hahn [8], and were supplemented by [9—17]. Further reviews of recent data are given in [18—19].

In 1974, Smith and Mitchell [20] reported on the determination of coalescence stability of paraffin oil-in-water emulsions with a preparative ultracentrifuge. They determined the minimum rotor speed required to produce a visible amount of coalesced oil after 30 min, and defined a "critical pressure" in this way. At the same meeting, Graham and Phillips [21] presented similar investigations of oil-in-water emulsions stabilized by proteins.

Using that method, Buscall [22] has determined coalescence pressures of O/W emulsions stabilized by a nonionic surfactant and surface-polymer mixtures, and has related the results to other physical-chemical measurements.

In our opinion, this critical pressure is a more suitable measure as it is equivalent to the forces determining the coalescence stability of emulsions.

The aim of our work was to examine whether it is possible to obtain coalescence pressures of emulsions using the centrifuge technique developed by El-Aasser and Robertson [23], Rohrsetzer, Kerek and Wolfram [24], and Melville, Willis and Smith [25] for solid particle dispersions (styrene butadiene latices, Berliner Blau sol, silver iodide). For this purpose, the following questions had to be answered: are final values reached after periods of centrifugation, i.e., is not the whole dispersed phase contained in the emulsion sample in the ultracentrifuge cell separated? How long will the final values remain constant, i.e., how does the centrifugal field itself influence the final values? Are the coalescence pressures calculated from the final values in agreement with the expected behavior of emulsion stability?

Experimental

Materials and preparation of the O/W emulsions

n-decane from "Petrolchemisches Kombinat Schwedt" (TGL 18691) and bi-distilled water were used. The continuous aqueous phase contained 0.1 mol dm^{-3} sodium dodecyl sulphate (SDS) (Merck, Darmstadt, for tenside tests) as stabilizer and no electrolyte.

This simple system was chosen because its physical-chemical properties have been studied in recent years,

both under standard conditions of temperature and pressure [18, 26—33] and centrifugal fields [6—8, 12, 15, 20].

The emulsions were prepared using a medical injection syringe (analogous to the advanced technique of Becher [34]) after adding 5.0 ml n-decane to 2.0 ml of the aqueous SDS-solution in a glass beaker by extracting and injecting through a capillary needle 10 times. By thinning with water and by the dye test it was proved whether the emulsions were of the oil-in-water type. The volume fraction of oil in the emulsions was about 71%. The droplet size, determined microscopically immediately after their formulation, ranged from about 1 to 30 µm.

Analytical ultracentrifugation

The emulsions were investigated immediately after their preparation or after various storage periods in a Model 3170 B Analytical Ultracentrifuge of MOM, Hungarian Optical Works, Budapest. The measurements were carried out at 20°C with 4° sector-shaped centerpieces of 12-mm thickness. The centrifugation technique has been described in detail in [2, 35].

Results and discussion

If the rotor speed is sufficient, e.g., 20000 rpm (30000 g, g = gravity acceleration) four layers can be observed from the bottom to the top of the cell (Fig. 1): a transparent continuous aqueous phase, a non-transparent emulsion, a transparent emulsion consisting of distorted oil droplets with thin aqueous films between them and a coalesced oil layer at the top. The transparent emulsion (often described as an oil-in-water biliquid foam [20, 36]) is also formed at lower rotor speeds if there is no oil separation. It disappears under gravity by penetration of continuous aqueous phase and becomes nontransparent again.

In similar emulsions (e.g., dodecane-water, SDS aq concentration 9×10^{-3} mol dm^{-3}, NaCl concentration 0.01 mol dm^{-3}, phase volume ratio oil/aq phase 1:1) the transparent emulsion remained also for a certain time after stopping the centrifugation, i.e., under gravity [20]. The reason for the different behavior is the difference in the electrolyte content influencing the thickness of the equilibrium films between emulsion droplets [29, 37]. In the case of the electrolyte-containing emulsion the water content in the transparent emulsion was found to be 0.1% and for about 10-µm dodecahedrons with uniform films between them a film thickness of only a few nm can be calculated [20]. Hoffmann et al.

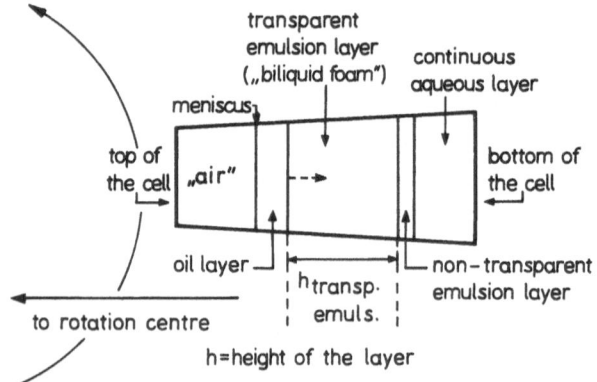

Fig. 1. Layers during demulsification of n-decane-in-water emulsions stabilized by 0.1 mol dm^{-3} aq SDS-soln. in an analytical ultracentrifuge

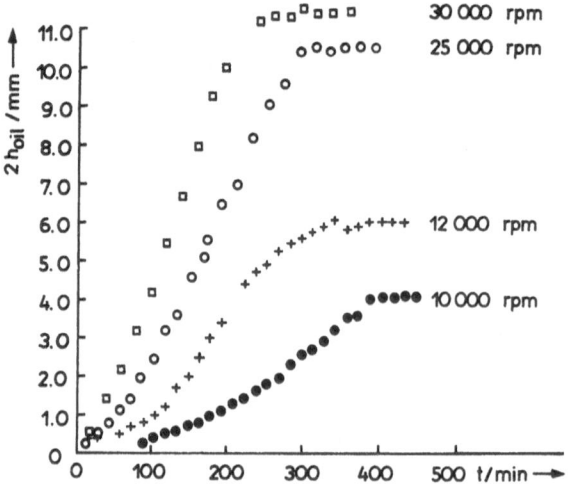

Fig. 2. Analytical ultracentrifugation of n-decane-in-water emulsions stabilized by 0.1 mol dm^{-3} aq SDS-soln. Dependence of the heights of separated oil layers on the time of centrifugation at various rotor speeds

[38] have obtained similar transparent concentrated emulsions by a quite different technique.

Depending on the centrifugation time, the oil/biliquid foam boundary moves to the bottom until constant heights of the oil and biliquid foam layers are obtained. Already during the acceleration period practically 100% of the continuous aqueous phase is separated. The non-transparent emulsion layer attains a constant height after a relatively short time of centrifugation.

In the reviewed literature there were no numerical data of final values, but there were a few curves showing constant heights of separated dispersed phase after certain periods of centrifugation, e.g., in [13]. On the other hand, we have found some indications of a complete demulsification of emulsions by centrifugation in [4, 13, 14]. In our investigations final values of the layers were established for all emulsions. Figure 2 shows the results of the analytical ultracentrifugation of n-decane-in-water emulsions stabilized by 0.1 mol dm^{-3} aq SDS-soln. at various rotor speeds. Final values are reached with all speeds after varying periods of centrifugation. At higher speeds there is a relatively long period of constant oil separation rate being the basis for the kinetic methods.

Figure 3 gives the results of stepwise raised rotor speeds, a technique also called "stepwise centrifugation". The emulsion sample was centrifuged at 20000 rpm (30000 g) until constant heights for all layers. Then the rotor speed was increased to 30000 rpm (65000 g), and it was centrifuged to final values of the layers again and so on. This technique is favorable as it is possible to obtain several final values of the same emulsion sample in the cell during one run.

A schematic presentation of the final values of all the layers shows that with increasing rotor speeds the oil/transparent emulsion boundary moves to the bottom (Fig. 4).

The sharp boundary between the transparent and non-transparent emulsion and the fact that the position of the boundary in the cell varies with the centrifugal force indicate a pressure barrier at the transition of the non-transparent to the transparent emulsion. The existence of such a pressure barrier has also been considered by Smith and Mitchell [20], but their experimental results show no movement of the transparent/non-transparent emulsion boundary in dependence on the rotor speed. Thus, they came to the conclusion that there was no pressure barrier. This result might be due to the preparative centrifuge employed, at which the reading of the positions of the boundaries is only possible after stopping the rotor. During braking of the rotor the transparent/non-transparent emulsion boundary again moves to the top, the faster the higher the applied centrifugal force. By applying an analytical ultracentrifuge, we were able to prove the movement of the transparent/non-transparent emulsion boundary in dependence on the relative centrifugal force with relatively high precision

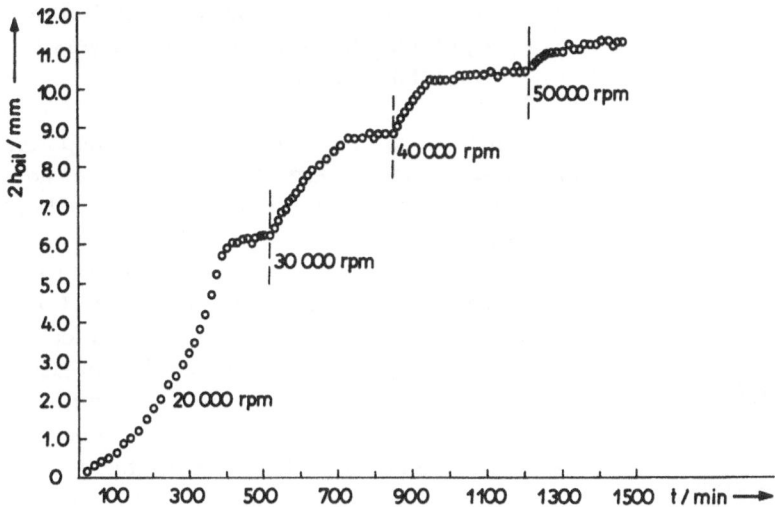

Fig. 3. Demulsification of a 0.1 mol dm⁻³ aq SDS-soln. stabilized emulsion by analytical ultracentrifugation. Dependence of the heights of n-decane layers on the centrifugation time in stepwise raised rotor speeds

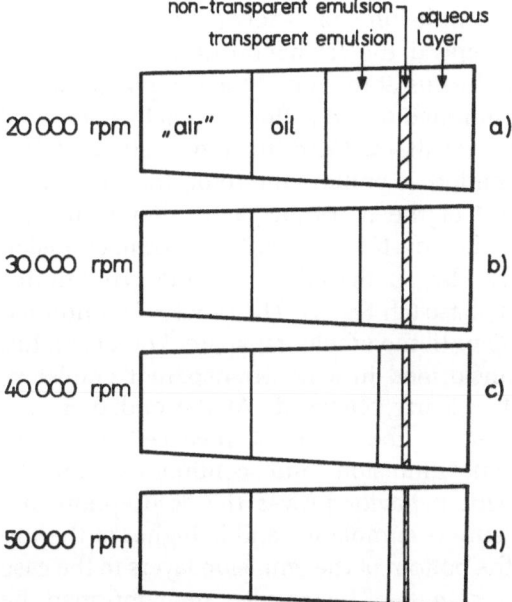

Fig. 4. Analytical ultracentrifugation of n-decane-in-water emulsions stabilized by 0.1 mol dm⁻³ aq SDS-soln. Change of the final values of layer heights in stepwise raised rotor speeds

Table 1. Analytical ultracentrifugation of a n-decane-in-water emulsion stabilized by 0.1 mol dm⁻³ aq SDS-soln. Dependence of the final values of the four layers on stepwise raised rotor speeds

Rotor speed/rpm (relative centrifugal force/g)		Final values of layer heights/mm			
		Oil	Transp. emulsion	Non-transp. emulsion	Aq. layer
20000	(30000)	3.09	2.95	0.51	2.44
30000	(65000)	4.47	1.78	0.29	2.42
40000	(120000)	5.29	1.00	0.22	2.41
50000	(200000)	5.65	0.67	0.15	2.37

(Table 1). This result has been confirmed by investigation of further SDS stabilized stock emulsions and also shows the advantage of the application of an analytical ultracentrifuge in such fields of research.

It is interesting that the emulsion sample in the cell is not demulsified completely at such high speeds

as 50000 rpm (200000 g). A small non-transparent and a transparent emulsion layer remain (Fig. 4 and Table 1). This could be due to the great stability of these emulsions; other explanations will be given below.

Using stepwise centrifugation up to a selected rotor speed the same final values were obtained compared with direct centrifugation at that speed, a result needed to compare the two types of centrifugation.

To obtain an answer to the question of how the centrifugal field itself influences the final values an emulsion was centrifuged at 30000 rpm for more than 20 h (Fig. 5). After a constant separation rate

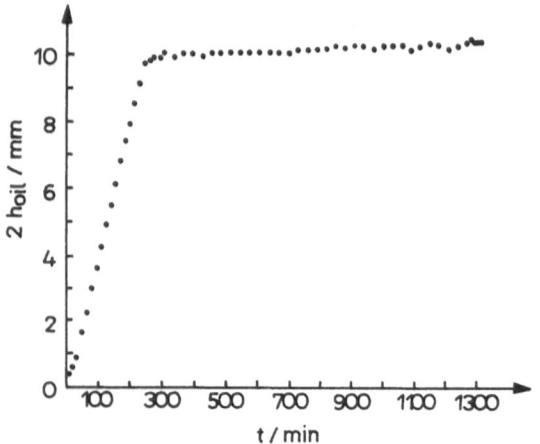

Fig. 5. Demulsification of a 0.1 mol dm^{-3} aq SDS-soln. stabilized n-decane-in-water emulsion by analytical ultracentrifugation at 30000 rpm (65000 g). Dependence of the heights of n-decane layers on the centrifugation time

of n-decane during the first 3 h a final value of oil layer height is attained and remains constatnt within the limits of error. It should be mentioned that such a good consistency has not been achieved in all emulsions with other oils and stabilizers. As a rule, the final values have been sufficiently constant at least for 1 to 3 hours.

Coalescence pressures

These results are the basis for deriving the pressure acting across the transparent + non-transparent emulsion layers as a measure of coalescence stability of emulsions under constant centrifugal fields. This pressure is also called "coalescence pressure" $P_{coal.}$. It is calculated for sector-shaped center-pieces as an approximation according to the Eq. (1)

$$ |\ P_{coal.}\ | = 1/2 \times \omega^2 \times \delta\rho \times \phi_{d.ph.}(r_1^2 - r_u^2)\ ,\ (1) $$

where

ω = angular velocity;
$\delta\rho$ = difference of densities of dispersed phase ($\rho_{d.ph.}$) and continuous phase ($\rho_{c.ph.}$);
$\phi_{d.ph.}$ = volume fraction of the dispersed phase in the emulsion layers;

r_1, r_u = distances from the lower (r_1) and upper (r_u) boundary of the emulsion layers to the rotation center.

The equation assumes that both the transparent and non-transparent emulsion layers are continuous. The force tending to bring about coalescence of the distorted droplets is considered to be cumulative from the bottom to the top of these layers (in the case of a negative buoyancy effect, i.e., $\rho_{d.ph.} < \rho_{c.ph.}$). As in the applied centrifugal technique, an equilibrium is attained, the pressure calculated at the oil/transparent emulsion boundary may be taken as the pressure required for coalescence. Similar equations have already been employed by Svedberg [39] for the partial hydrostatic pressure in gels and by other authors for latex dispersions [23], silver iodide sols [20], and O/W emulsions [21].

The investigation of the electrolyte-free SDS stabilized emulsions has shown that in increasing the rotor speed the formation of the transparent emulsion begins at the top of the sample in the cell and is continued to the bottom. If final values of all the layers are attained and the rotor is stopped during the period of braking the rotor, the increase of the height of the non-transparent emulsion layer begins at the non-transparent/transparent emulsion boundary by penetrating of the continuous aqueous phase (cf. Fig. 1). The process is continued to the top until the whole transparent emulsion has been transformed in a non-transparent emulsion, as can be clearly observed. At the end there are three layers in the ultracentrifuge cell: oil, non-transparent emulsion and continuous aqueous phase. This behavior proves the assumption that the pressure is cumulative and is higher at the top than at the bottom of the emulsion layers in the case of $\rho_{d.ph.} < \rho_{c.ph.}$. The results are confirmed by ultracentrifugation of latices with a negative buoyancy effect at which the particles were forced to the top and partly formed an irreversible coagulated film. Scanning electron microscopy of the two faces of the latex film showed that the coagulation is more advanced in the particles at the upper phase than in those at the lower [22].

Investigations of the dependence of the coalescence pressures on the storage periods of n-decane-in-water emulsions stabilized by 0.1 mol dm^{-3} aq SDS have indicated a marked increase of the pressures up to a storage period of 5 days at various rotor speeds (Table 2). This behavior is in

Table 2. Demulsification of 0.1 mol dm^{-3} aq SDS-soln. stabilized n-decane-in-water emulsions by analytical ultracentrifugation. Dependence of the coalescence pressures ($P_{coal.}$) on the storage periods of emulsions (St.p.-Em.)

Rotor speed: 30000 rpm (65000 g)		20000 rpm (30000 g)	
St.p.-Em./ days	$\mid P_{coal.} \mid$/ Pa $\times$ 10^5	St.p.-Em./ days	$\mid P_{coal.} \mid$/ Pa $\times$ 10^5
0	4.5	$^3/_4$	2.22
3	5.8	$1^3/_4$	2.45
4	6.4	$4^3/_4$	2.83
5	7.1	$5^3/_4$	2.85
11	7.0		

Table 3. Demulsification of n-decane-in-water emulsions by analytical ultracentrifugation. Dependence of the coalescence pressures ($P_{coal.}$) on the storage periods of the 0.1 mol dm^{-3} aqueous SDS-soln. (St.p.-SDS-soln.)

Rotor speed:	30000 rpm (65000 g)	
St.p.-SDS-soln/ days	31	177
$\mid P_{coal.} \mid$/Pa $\times$ 10^5	4.26	4.46

Table 4a)—c). "Stepwise analytical ultracentrifugation" of O/W emulsions of various particle sizes. Dependence of the coalescence pressures ($P_{coal.}$) on the rotor speed

Rotor speed/rpm (relative centri-fugal force/g)	$\mid P_{coal.} \mid$/ Pa $\times$ 10^5	

a) Particle size: 1—30 µm; oil: n-decane; stabilizer: 0.1 mol dm^{-3} aq SDS-soln.; I—III differ in storage periods of the SDS-soln. and emulsions

20000	(30000)	2.7	I
25000	(45000)	3.5	
15000	(17000)	2.3	II
20000		2.8	
20000		3.5	III
30000	(65000)	5.5	
40000	(120000)	7.0	
50000	(200000)	8.6	

b) Particles size: 0.8—2.6 µm, oil-concentrates: xylene/acetic acid ester, ethoxyl. alkyl phenol; storage periods oil conc. 8 (I), 27 (II) days

30000		1.01	I
32000	(75000)	1.04	
34000	(85000)	1.00	
35000	(90000)	0.95	
20000		0.73	II
25000		0.70	
30000		0.77	

c) Particle size: <1 µm, oil-conc.: n-decane/ethoxyl. fatty amine/nonyl phenol; St.p.-oil-conc. 16 month

10000	(7500)	0.19
20000		0.38
30000		0.73
50000		1.48

Particle size measured microscopically immediately after preparation of the emulsions

agreement with the well-known fact that some emulsions need a so-called "period of ripening" to attain a certain stability. During the first days after preparation of the emulsions chemical and/or physical processes take place which influence stability. The knowledge of such a behavior is important to predictions of emulsion stability.

Furthermore, a small increase of emulsion stability with increasing storage period of the aqueous SDS-solution has been found (Table 3). An explanation is that SDS hydrolyzes and the alcohol formed results in higher stability (cf. [33]).

The method has also been applied to food and herbicidal emulsions to investigate the influence of some parameters on the coalescence stability as well as to preparative laboratory centrifuges [40]. The results obtained are of interest with regard to practical purposes and in agreement with the expected behavior and other measurements.

Nevertheless, there remain some questions and problems to be solved, for example, the influence of particle size on the coalescence pressures. P_{coal} should be independent on the angular velocity for a monodisperse nondesorbing system [41]. Using stepwise centrifugation, we have found emulsions with no, small, and strong dependence of the coalescence pressures on the rotor speed. In the case of SDS containing polydisperse n-decane-in-water emulsions with particle sizes from about 1 to 30 µm a clear increase of coalescence pressures with increasing speeds has been obtained (Table 4a). One explanation for that could be the influence of particle size with the precondition that there is a stratification in particle size in the ultracentrifuge cell and that the smaller particles at the bottom give higher values of coalescence pressures than the larger ones at the top of the cell. In agreement with that, no influence of the rotor speed on the coalescence pressures in emulsions with particle sizes from about 1 to 3 µm has been revealed (Table 4b). On the other hand, in emulsions with particle sizes below 1 µm a strong influence of the speed on the coalescence pressures has been found (Table 4c). This means that the described behavior of the coalescence pressures cannot be explained by particle size of the emulsions alone.

Another explanation for the rise of the coalescence pressures with increasing rotor speeds could be that the additional amount of emulsifier which is set free by the breakdown of the emulsion stabilizes the remaining emulsion. Thus, higher pressures are required for a further separation of dispersed phase (see also [23]). The increase of the emulsifier concentration near the oil/transparent emulsion boundary is proportional to the amount of destroyed emulsion, i.e., the amount of oil. As this value depends on the rotor speed the increase of the calculated P_{coal} with the angular velocity is a clear implication for this hypothesis.

In such cases only small rotor speed steps should be chosen and the mean taken as an approximation for obtaining several final values during one run by stepwise centrifugation.

Conclusions

The analytical ultracentrifuge is a useful tool for determining the coalescence stability of emulsions. Some investigations with monodisperse emulsions are necessary for better understanding of the phenomena concerning the dependence of the coalescence pressures on the particle size.

References

1. Strenge K, Seifert A (1984, 10.—15. Dezember) II. Informationssymp. Kolloidwissenschaften, Kühlungsborn, FRG
2. Seifert A (1988, April) Experiment. Studie Zentralinst Physikal Chem, Berlin-Adlershof, FRG
3. Merill RC (1943) Ind Eng Chem Anal ed 15:743—746
4. Rehfeld SJ (1962) J Phys Chem 66:1966—1968
5. Garrett ER (1962) J Pharm Sci 51:35—42
6. Vold RD, Groot RC (1962) J Phys Chem 66:1969—1975
7. Hahn AU, Vold RD (1975) J Colloid Interface Sci 51:133—142
8. Vold RD, Mittal KL, Hahn AU (1978) In: Matijevic E (ed) Surface Colloid Sci Ser Vol 10 Plenum Press, New York, pp 45—97
9. Bogs U, Naumann H (1963) Pharmazie 18:750—754
10. Sharma MK (1977) Acta Cienc Indica 3:40—42, 139—141; Current Sci (India) 46:131—132; Indian J Chem Sect. A 15A:644—645; Sci Cult 43:456—458
11. Okamoto K, Oishi H (1977) Yakugaku Zasshi 97:251—256
12. Vold RD, Malectic M (1978) J Colloid Interface Sci 65:390—393; Proc VII. Int Congress Surf Act Subst (1976) (Pub. 1978) Sect B, Pt I, Vol 2, Nats Kom SSSR Poverkhn-Akt Veshchestvam: Moscow, USSR, pp 608—614
13. Hahn AU, Mittal KL (1979) Colloid Polym Sci 257:959—967
14. Larichev NA, Gurov AN, Tolstoguzov VB (1983) Colloids Surf 6:27—34
15. Vold MJ (1985) Langmuir 1:74—78
16. Ostrovsky MV, Good RJ (1986) J Dispersion Sci Technol 7:95—125
17. Kedvessy G, Erös I, Mednyasnzky A, Morosz M (1987) Pharm Ind 49:747—751
18. Tadros TF, Vincent B (1983) In: Becher P (ed) Encyclopedia of Emulsion Technology Vol 1, Marcel Dekker, New York and Basel, pp 129—285; Menon VB, Wasan TD (1985) ibid 2:1—75
19. Seifert A, Strenge K, Sonntag H (1983) Literaturstudie Zentralinst Physikal Chem, Berlin-Adlershof, FRG
20. Smith AL, Mitchell DP (1976) In: Smith AL (ed) Theory and practice of emulsion technology (Proc Symp 1974) Academic Press London, New York, pp 61—74
21. Graham DE, Phillips MC, ibid., pp 75—98
22. Buscall R (1978) Progr Colloid Polym Sci 63:15—26
23. El-Aasser MS, Robertson AA (1971) J Colloid Interface Sci 36:86—93
24. Rohrsetzer S, Kerek I, Wolfram E (1971) Kolloid-Z Z Polym 245:529—530
25. Melville JB, Willis E, Smith AL (1972) J Chem Soc Faraday Trans 1, 68:450—455
26. Van den Tempel M (1957) In: Schulman JH (ed) Gas/liquid and liquid/liquid interface (Proc 2nd Internat Congr Surf Activity, I) Butterworth, London, pp 439—446
27. Rowe EL (1965) J Pharm Sci 54:260—264
28. Jones MN, Mysels KJ, Scholten PC (1966) Trans Faraday Soc 62:1336
29. Sonntag H, Strenge K (1972) Coagulation and stability of disperse systems, Halstedt Press, New York, pp 64—67
30. Davis SS, Smith A (1972) J Pharm Pharmac 24 (Suppl): 155P—156P
31. Davis SS, Smith A (1973) Kolloid-Z Z Polym 251:337—342

32. Princen HM, Aronson MP, Moser JC (1980) J Colloid Interface Sci 75:246—270
33. Mysels KJ (1986) Langmuir 2:423—428
34. Becher P (1967) J Colloid Interface Sci 24:91—96
35. DD-WP 274674
36. Sebba F (1972) J Colloid Interface Sci 40:468—474
37. Sonntag H, Netzel J (1972) Z phys Chem (Leipzig) 250:119—123
38. Hoffmann H, Ebert G (1988) Angew Chem 100:933—944
39. Svedberg T, Pedersen KO (1940) Die Ultrazentrifuge, Theodor Steinkopff, Dresden und Leipzig 1940, pp 26—29
40. in preparation
41. Strenge K (1983) Plaste Kautschuk 30:450—451

Authors' address:

Dr. A. Seifert
Zentralinstitut für Ernährung
AG Lebensmittelchemie
Arthur-Scheunert-Allee 114—116
O-1505 Bergholz-Rehbrücke, FRG

Progress in Colloid & Polymer Science   Progr Colloid Polym Sci 86:84—91 (1991)

Swelling pressure equilibrium of swollen crosslinked systems in an external field. I: Theory

W. Borchard

Angewandte Physikalische Chemie der Universität-GH-Duisburg, FRG

Abstract: The deformation of an elastic fluid mixture in a centrifugal field leading to a continuous equilibrium has been described. The binary gel is assumed to remain isotropic in the deformed state. As soon as a boundary gel/solvent appears an osmotic pressure identical to the swelling pressure can be calculated which depends on the concentration inside the gel. For highly swollen gels closed expressions for the Svedberg-Pedersen equation are obtained. The influence of the hydrostatic pressure on the heterogeneous swelling equilibrium at the boundary gel/solvent has also been considered.

Key words: Gels; deformation; ultracentrifugal field; swelling pressure; discontinuous/continuous equilibrium

Introduction

Chemically or physically crosslinked substances may take up solvent and form gels. These systems are elastic or viscoelastic liquid mixtures which may undergo well known heterogeneous equilibria with the pure solvent or solutions. If at isothermal conditions both phases are coexisting under the same pressure, this is the well-known swelling equilibrium [1—7]. The case of the pressure in the gel phase differing from that of the solvent phase is called the osmotic swelling equilibrium or swelling pressure equilibrium [3, 4, 8—13].

Long ago, in the nearly unknown paper of Riecke, the anisotropic deformation of a swollen crosslinked system in coexistence with the solvent was treated [1]. As long as the stresses in the deformed body and the corresponding deformations do not vary in a certain direction in the gel, the concentration will also be the same along this direction. This is generally the case if the forces are acting on the surface of a body. We will show that the composition is depending on the position inside the gel if external forces like gravitational or centrifugal forces are operating. The case of the deformation of gels in an ultracentrifugal field has been considered under restricting conditions by Svedberg and Pedersen for the first time [14].

A combined treatment using general thermodynamics and the statistical theory of Flory [15] has been presented by Bloomfield [16]. In the model calculations a rectangular piece of a gel is deformed anisotropically in a centrifugal field. It has been shown that the deswelling effect due to the field is most distinct if the solvent/polymer system is close to demixing.

Only a few experiments on the sedimentation of gels have been performed in a centrifugal field [14, 20—28].

In the following, we want to make the assumption that the gel phase remains isotropic during the deformation, which is really not the case if the deformations or stresses are large [17]. This implies that the stress tensor can be represented by its diagonal terms [6, 18]. Further, the system is considered to be a binary one where the solvent and the crosslinked polymer are the components. Both species are assumed to be in the state of internal equilibrium.

We illustrate the deformation of the gel in a centrifugal field by means of Fig. 1.

At time $t = 0$ it is supposed that the cell is brought up to the constant angular velocity ω. A movement of the meniscus gel/vapor at distance $r_m^{g/v}$ due to sedimentation and diffusion has not yet taken place. This is shown in part a). If the gel (g) is in

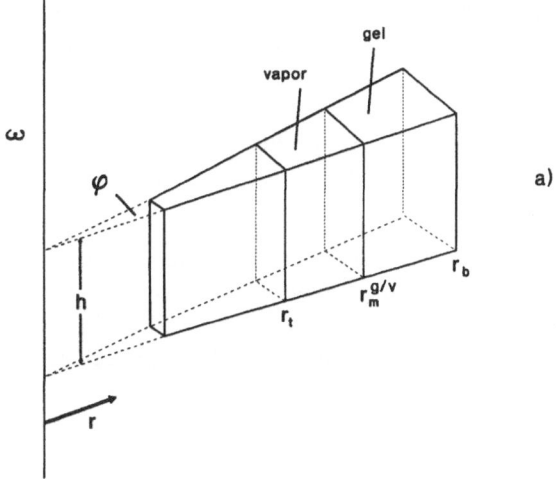

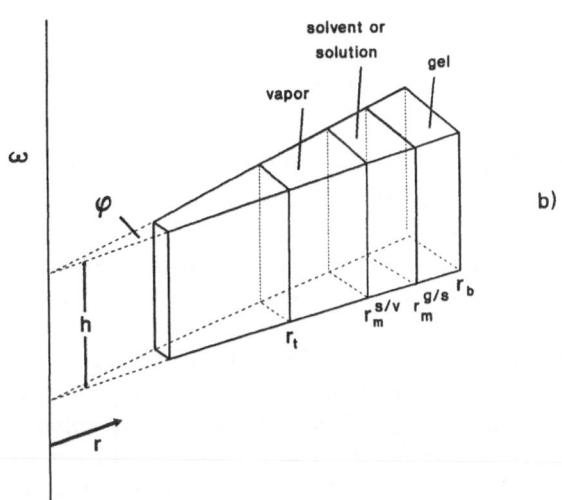

Fig. 1. Schematic representation of the deformation of a gel in an analytical centrifuge, r = radial distance from the axis of centrifugation, ω = angular velocity, φ = sector angle of the center piece containing the gel, h = height of the center piece, indices: t = top of the cell, m = meniscus, b = bottom of the cell;
phase boundaries: g/v = gel/vapor, s/v = solvent/vapor, g/s = gel/solvent. a) at time $t = 0$ the gel is situated between $r_m^{g/v}$ and r_b; b) deformation of the gel from $r_m^{g/v}$ to $r_m^{g/s}$ in equilibrium (see text)

its saturation state of swelling we suppose that it is coexisting with an infinitely small layer of solvent (s) or solution on its surface with the phase boundaries g/s and s/v facing the gel and the vapor phase (v), so that $r_m^{g/v}$ and $r_m^{s/v}$ are practically the same. The pressure at this distance is given by the hydrostatic

pressure of the vapor phase which is approximately 1 atm, therefore, the swelling is nearly the same as that measured outside the centrifugal field. At the bottom of the cell and for all distances except the original position $r_m^{g/v}$ the osmotically active pressure changes so that the degree of swelling will change also. It will take a certain time until the equilibrium state b) is reached. The consequence of the change in pressure is that also the meniscus g/s moves to the distance $r_m^{g/s}$. At this position the gel surface is under the hydrostatic pressures of the solvent or solution and the vapor phase at the same distance which is mainly given by the hydrostatic pressure of the solvent or solution phase. The swelling of the gel at its surface at the meniscus g/s depends on this pressure as it is known from the equilibrium description [3]. In a cell which has the shape of a sector of a rotating cylinder the surface layer of the gel is slightly stretched in the tangent direction perpendicular to the radial and axial direction when it moves into the direction of the bottom of the cell. This effect is small for small deformations and corresponds to the dilution effect known from the centrifugation of solutions [19]. It will lead to a slight anisotropy which is ignored in this treatment.

Up to now, we have said that the gel in its original state is swollen in a solvent or a solution. The latter may occur if the swelling agent itself is a mixture either of different low molecular solvents or of a low molecular solvent and an oligomer or polymer substance. As we restrict ourselves only to a binary gel, we have to add that the solvent is considered to be a one-component phase, although we know that very complicated phenomena may occur if these conditions are violated [20, 21].

Theory

A) The heterogeneous and continuous swelling pressure equilibrium

A gel in an external field is a continuous system which is governed by the gradient of the potential of the external field under isothermal conditions. The generalized specific force $\bar{X}_i$ which is the force per mass of component i is related to the gradient of the specific potential of the component i of the external field $\bar{\phi}_i$ and of the specific chemical potential $\bar{\mu}_i$ by Eq. (1), where, in the rotating system with one axis of rotation, only the radial distance from this axis r is of importance

$$\tilde{X}_i = -\frac{d\tilde{\phi}_i}{dr} - \frac{d\tilde{\mu}_i}{dr} \; ; \quad i = 1,2 \; . \tag{1}$$

If at time $t = 0$ a homogeneously swollen gel is placed into a centrifugal field the generalized force $\tilde{X}_i$ is given at this instant by the first term in Eq. (1). But by the action of the external field according to deformation and diffusion a concentration gradient will be built up inside the gel until the gradient of the chemical potential of i has just been equalized by that of the external field. This corresponds to the equilibrium of a continuous system in which the generalized driving force has to vanish.

For a centrifugal field the first term of Eq. (1) is given by Eq. (2)

$$-\frac{d\tilde{\phi}_i}{dr} = \omega^2 r \; , \tag{2}$$

where ω is the angular velocity. The righthand side of Eq. (2) is the centrifugal acceleration. This has to be substituted by the gravitational acceleration if the gel is exerted to a gravitational field.

As we are only considering isotropic gels, the specific chemical potential of i depends on temperature T, pressure P, and the mass concentration or partial density of i given by ρ_i, which can be written in the form

$$\bar{\mu}_i = \bar{\mu}_i(T, P, \rho_i) \; . \tag{3}$$

The total differential of this equation divided by dr leads to

$$\frac{d\bar{\mu}_i}{dr} = \tilde{V}_i \frac{dP}{dr} + \left(\frac{\partial \bar{\mu}_i}{\partial \rho_i}\right)_{T,P} \frac{d\rho_i}{dr} \; . \tag{4}$$

$\tilde{V}_i$ is the partial specific volume of component i.

The equilibrium condition already mentioned is formulated by Eq. (5)

$$\tilde{X}_i = 0 \; . \tag{5}$$

By use of Eqs. (1), (2), (4) and (5), we get after multiplication by ρ_i,

$$\rho_i \omega^2 r = \rho_i \tilde{V}_i \frac{dP}{dr} + \rho_i \left(\frac{\partial \bar{\mu}_i}{\partial \rho_i}\right)_{T,P} \frac{d\rho_i}{dr} \; . \tag{6}$$

Summing up Eq. (6) for $i = 1$ and $i = 2$ and considering that the density of the gel ρ is given by $\rho = \rho_1 + \rho_2$, we obtain

$$\rho \omega^2 r = \frac{dP}{dr} + \sum_{i=1}^{2} \rho_i \left(\frac{\partial \bar{\mu}_i}{\partial \rho_i}\right)_{T,P} \frac{d\rho_i}{dr} \; . \tag{7}$$

The Gibbs-Duhem relation tells us that the sum in Eq. (7) is zero for a given value r. This leads to the condition for the mechanical equilibrium in a continuous system, as in the case of a solution [17]

$$\rho \omega^2 r = \frac{dP}{dr} \; . \tag{8}$$

From Eqs. (6) and (8) we deduce

$$\omega^2 r (1 - \tilde{V}_i \rho) dr = \left(\frac{\partial \bar{\mu}_i}{\partial \rho_i}\right)_{T,P} d\rho_i \; . \tag{9}$$

The lefthand side of Eq. (9) contains the buoyancy term in brackets, the righthand side describes the change of the concentration for component $i = 1$ or $i = 2$. There will be no concentration gradient in the gel phase if either ω or the buoyancy term is equal to zero.

The deformation of the gel by a body force generates an osmotic pressure or swelling pressure because it acts in a different way on the solvent, e.g., $i = 1$ and the gel $i = 2$. Solvent and gel are allowed to move freely to the equilibrium positions, which would not be the case in a simple compression experiment, where each shell of the gel normal to the direction of r has to correspond to a closed system.

The difference between an osmotic pressure in an open system and the compression of a closed system has been demonstrated in Fig. 2 schematically for a uniform static pressure. In the centrifugal field the osmotically operating pressure is dependent on the radial position in the gel. Svedberg proposed the name "partial hydrostatic pressure of the solvent", which is misleading. We prefer "osmotically active pressure", which corresponds to the swelling pressure in case of an equilibrium sedimentation of the gel phase. The only difference in comparison to the well known swelling pressure equipment is that, in a centrifugal field, the swelling pressure-concentration curve starting from the value zero can be obtained in a

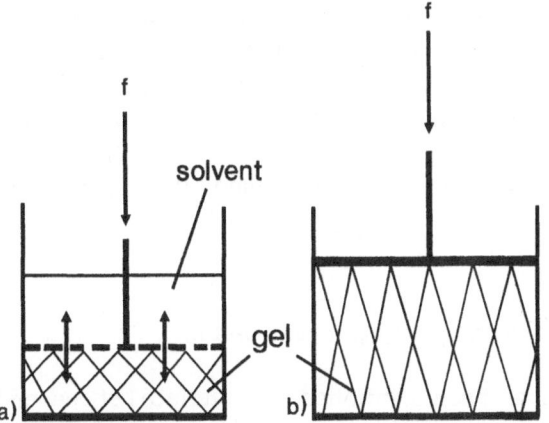

Fig. 2. Schematic representation of a gel enclosed in a cylinder and a piston; a) force f is acting on a piston permeable to the solvent, but not permeable to the gel, b) the piston is not permeable to the solvent or the gel, the double arrow indicates the permeability of the solvent (see text)

single equilibrium run of the centrifuge [14, 20, 21, 25–28].

From these considerations it follows that we have to treat the equilibrium as being continuous inside the gel phase, but discontinuous at the meniscus gel/solvent or solution, which is clearly in variance with the sedimentation-diffusion equilibrium known from polymer solutions, which is continuous if we exclude demixing [29].

For the gel at the phase boundary g/v at time $t = 0$, and only if the mentioned swelling equilibrium of a binary gel has been established, we have the condition

$$\bar{\mu}_{01}(T, P_0) = \bar{\mu}_1^g(T, P_0, \rho_1) \,, \tag{10}$$

where P_0 is the atmospheric pressure, T is the temperature, $\bar{\mu}_1^g$ the chemical potential of the solvent in the gel phase (index g), and ρ_1 the partial density of the solvent at saturation. In the case in which this boundary moves from $r_m^{g/v}$ to the equilibrium position $r_m^{g/s}$ with the pressure P_{ref}, this equation reads

$$\bar{\mu}_{01}(T, P_{ref}) = \bar{\mu}_1^g(T, P_{ref}, \rho_1) \,. \tag{11}$$

The case in which a gel has not yet reached the swelling maximum will be discussed later. We use the indication reference because, starting from this pressure, the swelling pressure will be calculated.

For the continuous gel Eq. (9) can be written in the form

$$\rho_i \omega^2 r (1 - \tilde{V}_i \rho) dr = \rho_i \left(\frac{\partial \bar{\mu}_i}{\partial \rho_i} \right)_{T,P} d\rho_i$$

$$= \rho_i (d\bar{\mu}_i)_{T,P} \,. \tag{12}$$

The differential $(d\bar{\mu}_i)_{T,P}$ is the total change of the specific chemical potential of component i at constant values of T and P. Using the Gibbs-Duhem relation for constant values of T and P:

$$\rho_1 (d\bar{\mu}_1)_{T,P} + \rho_2 (d\bar{\mu}_2)_{T,P} = 0 \,, \tag{13}$$

we obtain with Eq. (12) for $i = 1$ and $i = 2$

$$\rho_1 \omega^2 r (1 - \tilde{V}_1 \rho) dr = \rho_1 (d\bar{\mu}_1)_{T,P}$$

$$= -\rho_2 (d\bar{\mu}_2)_{T,P} \,, \tag{14a}$$

and

$$\rho_2 \omega^2 r (1 - \tilde{V}_2 \rho) dr = \rho_2 (d\bar{\mu}_2)_{T,P}$$

$$= -\rho_1 (d\bar{\mu}_1)_{T,P} \,, \tag{14b}$$

where we see that the sum of Eqs. (14a) and (14b) equals zero on both sides. With these equations we have the choice of calculating the swelling pressure in the gel phase via the change of the chemical potential of the solvent given by Eq. (14a) or (14b). Starting from the definition of the osmotic pressure difference in a homogeneous system coexisting with the pure solvent, we have for the difference of the specific chemical potential [30]

$$-\int_{P_{ref}}^{P} \tilde{V}_1 dP \equiv \Delta\bar{\mu}_1$$

$$= \bar{\mu}_1(T, P_{ref}, \rho_1) - \bar{\mu}_{01}(T, P_{ref}) \,. \tag{15}$$

P_{ref} is a reference pressure at the gel's surface adjacent to the solvent, which at the beginning of the experiment is identical to the atmospheric pressure P_0. The osmotically active pressure difference ΔP_{os} is given by

$$P - P_{ref} \equiv \Delta P_{os} \,. \tag{16a}$$

This pressure difference is a differential swelling pressure and is identical to the swelling pressure as soon as the equilibrium concentration gradient has been built up in the gel and the swelling equilibrium has been reached at the phase boundary between gel and solvent:

$$\Delta P_{os} = \Pi_s \ . \tag{16b}$$

By integration of Eq. (14a), we get, by using Eq. (15),

$$\Delta \bar{\mu}_1 = -\int_{P_{ref}}^{P} \tilde{V}_1 dP = \omega^2 \int_{r_{ref}}^{r} r(1 - \tilde{V}_1 \rho) dr \ . \tag{17}$$

The distance r_{ref} corresponds to P_{ref} at the gel boundary g/s. For an incompressible gel $\tilde{V}_1$ is independent of the pressure. This leads to the approximation

$$-\omega^2 \int_{r_{ref}}^{r} r(1 - \tilde{V}_1 \rho) dr = \int_{P_{ref}}^{P} \tilde{V}_1 dP \approx \tilde{V}_1 \Delta P_{os} \ . \tag{18a}$$

With the approximation $\tilde{V}_1 \approx \tilde{V}_{01}$, which means that the partial specific volume of the solvent can be replaced by the specific volume of the pure solvent $\tilde{V}_{01}$, it follows that

$$-\omega^2 \int_{r_{ref}}^{r} r(1 - \tilde{V}_{01} \rho) dr \approx \tilde{V}_{01} \Delta P_{os} \ . \tag{18b}$$

In the same way, we get from Eqs. (14b), (15), and (16):

$$\omega^2 \int_{r_{ref}}^{r} \frac{\rho_2}{\rho_1} r(1 - \tilde{V}_2 \rho) dr = \int_{P_{ref}}^{P} \tilde{V}_1 dP \approx \tilde{V}_1 \Delta P_{os} \ , \tag{19a}$$

and

$$\omega^2 \int_{r_{ref}}^{r} \frac{\rho_2}{\rho_1} r(1 - \tilde{V}_{02} \rho) dr \approx \tilde{V}_{01} \Delta P_{os} \ . \tag{19b}$$

As the righthand sides of Eqs. (18a) and (19a), and also of Eqs. (18b) and (19b) are the same, we can use either the integrals on the lefthand sides of Eq. (18a) resp. (19a), or Eq. (18b) resp. (19b) for the evaluation of the osmotically active pressure difference respective the swelling pressure.

In any case, the density of the gel ρ and ρ_2 or ρ_1 as a function of the axial distance r have to be known, which can be calculated from the original overall composition of the gel and, for instance, the experimentally available Schlieren pattern. Equation (18a) or (19a) is called the generalized Svedberg-Pedersen equation [31].

B) Influence of the hydrostatic pressure

As has been mentioned, the interface between gel and solvent g/s is under a hydrostatic pressure which is dependent on the height of the liquid column given by the difference of $r_m^{g/s}$ and $r_m^{s/v}$. From Eq. (8), it follows for the solvent phase,

$$dP = \rho_{01}(r) \omega^2 r dr \ , \tag{20}$$

where $\rho_{01}(r)$ is the density of the pure solvent which normally depends on r and thereby on P. Integration leads to

$$\Delta P = P_{ref} - P_0 = \int_{r_m^{s/v}}^{r_m^{g/s}} \rho_{01}(r) \omega^2 r dr \ , \tag{21}$$

which will give in the case of a constant density for small column heights, approximately,

$$\Delta P \approx \omega^2 \rho_{01} \frac{(r_m^{g/s})^2 - (r_m^{s/v})^2}{2} \ . \tag{21a}$$

Thus, the pressure at the meniscus gel/solvent which is P_{ref} is given by

$$P_{ref} = P_0 + \omega^2 \rho_{01} \frac{(r_m^{g/s})^2 - (r_m^{s/v})^2}{2} \ . \tag{21b}$$

This hydrostatic pressure may change the degree of swelling of the gel which is described by means of Eq. (11). Under isothermal conditions and using the partial density of the solvent as the variable for concentration the total differential of Eq. (11) results in

$$\left(\frac{\partial \bar{\mu}_1^g}{\partial \rho_1} \right)_{T,P} d\rho_1 + \tilde{V}_1^g dP = \tilde{V}_{01} dP \ . \tag{22}$$

The first term in brackets refers to the maximum of swelling. Eq. (22) first has been used by Rehage to describe the pressure dependence of the swelling equilibrium where a different concentration variable has been used [3].

The combination of Eqs. (20) and (22) leads to

$$d\rho_1 = \left[(\tilde{V}_{01} - \tilde{V}_1^g) / \left(\frac{\partial \tilde{\mu}_1^g}{\partial \rho_1} \right)_{T,P} \right] \rho_{01}(r)\omega^2 r\, dr \ . \quad (23)$$

This can be integrated between P_0 and P_{ref}, respectively, between the quantities $r_m^{g/s}$ and $r_m^{s/v}$

$$\int_{P_0}^{P_{\text{ref}}} d\rho_1 = (\rho_1)_{P_{\text{ref}}} - (\rho_1)_{P_0}$$

$$= \omega^2 \int_{r_m^{s/v}}^{r_m^{g/s}} \left[(1 - \tilde{V}_1^g \rho_{01}) / \left(\frac{\partial \tilde{\mu}_1}{\partial \rho_1} \right)_{T,P} \right] r\, dr \ . \quad (24)$$

Finally, we obtain for the partial densities at the reference pressure P_{ref} and atmospheric pressure P_0 the relation

$$(\rho_1)_{P_{\text{ref}}} = (\rho_1)_{P_0}$$

$$+ \omega^2 \int_{r_m^{s/v}}^{r_m^{g/s}} \left[(1 - \tilde{V}_1^g \rho_{01}) / \left(\frac{\partial \tilde{\mu}_1}{\partial \rho_1} \right)_{T,P} \right] r\, dr. \quad (24)$$

From the stability conditions it is known that $(\partial \tilde{\mu}_1 / \partial \rho_1)_{T,P} > 0$ holds for stable and metastable phases. The integral vanishes if $\tilde{V}_1^g = \tilde{V}_{01}$ which means that the partial specific volume of the solvent in the gel is the same as the specific volume of the pure solvent. If this is valid for all concentrations it corresponds to the additivity of the volume of the gel with respect to the volumes of the pure components. Thus, we are able to calculate the concentration at the meniscus gel/solvent if $\tilde{V}_1^g$ and $(\partial \tilde{\mu}_1 / \partial \rho_1)_{T,P}$ are known, although in some systems the effect is expected to be small.

Discussion

It is easily shown that the results of Eqs. (18a) and (19a) are completely identical, which results from the equivalence of the terms

$$-(1 - \tilde{V}_1 \rho) = (1 - \tilde{V}_2 \rho) \frac{\rho_2}{\rho_1} \ ,$$

because $\tilde{V}_1 \rho_1 + \tilde{V}_2 \rho_2 = 1$ and $\rho = \rho_1 + \rho_2$. We call Eq. (19a) the generalized Svedberg-Pedersen equation, because it is general under the conditions mentioned and leads to the relation obtained by Svedberg and Pedersen if the conditions for highly swollen gels are introduced. From Eq. (19a), we get immediately

$$\Delta P_{\text{os}} = \omega^2 \int_{r_{\text{ref}}}^{r} \frac{\rho_2}{\tilde{V}_1 \rho_1} (1 - \tilde{V}_2 \rho) r\, dr \ . \quad (19a)$$

With the approximations $\rho_2 \ll \rho_1$ and, therefore, a) $\tilde{V}_1 \rho_1 \approx 1$, b) $\rho \approx \rho_{01}$, and c) $\tilde{V}_2 \approx \tilde{V}_{02}$, where $\tilde{V}_{02}$ is the specific volume of pure component 2, we get

$$\Delta P_{\text{os}} = \omega^2 \int_{r_{\text{ref}}}^{r} \rho_2 (1 - \tilde{V}_{02} \rho_{01}) r\, dr \ , \quad (19c)$$

which is the result of Svedberg and Pedersen [14]. The conditions a), b), and c) mean that the volume of the polymer can be neglected, the density of the gel is given by the partial density of the solvent, and that the partial specific volume of component 2 can be replaced by the specific volume of the polymer. It is clear that these approximations are only valid for extremely high degrees of swelling of the gel. If higher concentrated gels are investigated it is necessary to start from Eq. (18a) or Eq. (19a) in order to keep the error as small as possible.

In some cases, we used the approximation a) following Eq. (19a), which results in

$$\Delta P_{\text{os}} = \omega^2 \int_{r_{\text{ref}}}^{r} \rho_2 (1 - \tilde{V}_2 \rho) r\, dr \ . \quad (19d)$$

This will result in an error of 3% when gels of up to 10% by weight are investigated [20, 27]. The most useful relation is obtained from Eq. (18a), which results in

$$\Delta P_{\text{os}} = \omega^2 \int_{r_{\text{ref}}}^{r} \left(\rho - \frac{1}{\tilde{V}_1} \right) r\, dr$$

$$\approx \omega^2 \int_{r_{\text{ref}}}^{r} \left(\rho - \frac{1}{\tilde{V}_{01}} \right) r\, dr \ . \quad (19e)$$

In the above treatment, we assumed that the initial concentration of the polymer in the gel ρ_2 corresponds to that of the swelling equilibrium at atmospheric pressure, which we will name $\rho_{2,s}$. The pressure difference ΔP_{os} equals the swelling pressure Π_s.

Thermoreversible gels which are formed by a physical reaction of the polymer like a coil-helix transition with subsequent aggregation of the helices may correspond to a state of swelling where $\rho_2 \neq \rho_{2,s}$.

If the actual polymer concentration ρ_2 is higher than $\rho_{2,s}$ at the beginning of the centrifugal experiment the osmotically active pressure difference ΔP_{os} will cause a concentration gradient in the gel phase as it is schematically demonstrated in Fig. 3 for a constant rotational speed ω_1.

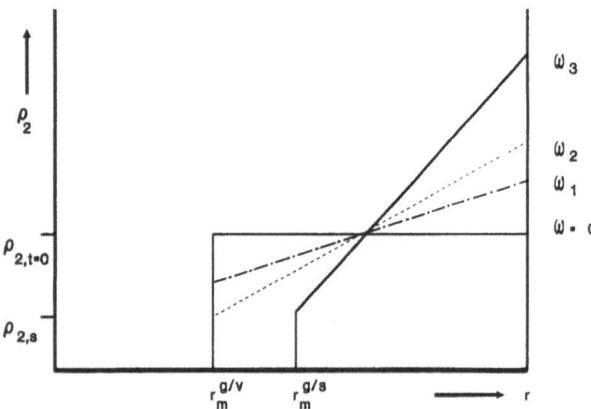

Fig. 3. Schematic representation of concentration profiles inside a gel for different rotational speeds ω_i. ρ_2 = partial density of the polymer; the indices $t = 0$ and s correspond to zero time and saturation at swelling equilibrium. At time $t = 0$ the concentration of component 2 is higher than $\rho_{2,s}$; $r_m^{g/v}$ and $r_m^{g/s}$ are the radial distances between meniscus m and rotational axis for the phase boundaries gel/vapor (g/v) and gel/solvent (g/s) (see text)

If ω is increased the equilibrium concentration at the bottom of the cell increases further and that at the boundary gel/vapor decreases. As the concentration at the phase boundary cannot drop below $\rho_{2,s}$, the meniscus of the gel has to move towards the bottom for rotational speeds $\omega > \omega_2$. The latter has been assumed to correspond to a necessary speed so that $\rho_2 = \rho_{2,s}$ at the boundary. For $\omega_3 > \omega_2$ the gel boundary has moved from the original

position $r_m^{g/v}$ to $r_m^{g/s}$. From these considerations it follows that a gel boundary can only shift for rotational speeds being higher than a critical one. For gels swollen to a maximum value $\rho_{1,s}$ or minimum value $\rho_{2,s}$ the gel boundary will shift as soon as an external field is applied. The osmotically active pressure difference corresponds to the swelling pressure given by Eq. (16b). It has to be mentioned that the concentration $\rho_{2,s}$ at the meniscus gel/solvent has to be known for the integration, for example, of Eq. (19e).

In the case of syneresis possible physical reactions may increase the crosslinking density. In this case the partial density of the solvent is a function of time $\rho_1 = \rho_1(t)$ [32, 33, 35]. Therefore, sedimentation and physical reaction are superimposed. If a strong syneresis leads to an exudation of the solvent at early stages of the sedimentation complicated concentration profiles in the gel may occur.

In some cases it may be difficult to measure the concentration gradient in the gel phase because of a poor transparency of the fluid elastic mixture. Then the concentration gradient inside the gel can be calculated by application of the mass balance between $r_m^{g/s}$ and r_b if the concentration gradient is known to be constant or to change with the distance r monotoneously [14, 20, 21, 25—28].

Comparing the brackets of Eqs. (18a) and (24), we can clearly see the difference between an osmotic and a hydrostatic pressure, both provoked by the centrifugal acceleration $\omega^2 r$.

In Eq. (24) the term $(\partial \tilde{\mu}_1 / \partial \rho_1)_{T,P}$ enters in the denominator, which is zero at the stability limit of a gel. Therefore, a large change of the degree of swelling with pressure at the gel surface $r_m^{g/s}$ is only expected if the swelling curves will show a nearly horizontal course in the temperature-concentration curve with an inflexion point for $\tilde{V}_1^s \neq \tilde{V}_{01}$ [34]. This is in agreement with the model calculations of Bloomfield [16].

Conclusion

Closed expressions for the generalized Svedberg-Pedersen equation have been derived which describe the behavior of homogeneous fluid elastic mixtures in an ultracentrifugal field. The method can be applied to systems of high polymer concentrations and, generally, will result in a swelling pressure-concentration curve in a range of concentrations. With these thermodynamic data it is possible to test swelling theories for crosslinked swollen systems.

Acknowledgement

We thank the Deutsche Forschungsgemeinschaft and the Ministerium für Wissenschaft und Forschung des Landes NRW for financial support of this project.

References

1. Riecke E (1894) Wied Ann 53:564
2. Breitenbach JW, Frank HP (1948) Mh Chemie 79:531
3. Rehage G (1964) Kolloid-Z u Z Polymere 194:16
4. Rehage G (1964) Kolloid-Z u Z Polymere 196:97
5. Rehage G (1964) Kolloid-Z u Z Polymere 199:1
6. Borchard W (1975) Habilitationsschrift, Clausthal
7. Borchard W, Steinbrecht U (1991) Colloid Polym Sci 269:95
8. Borchard W (1966) Dissertation, Aachen
9. Posnjak E, Freundlich H (1912) Kolloid-Beih 3:417
10. Prins W, Pennings AJ (1961) J Polym Sci 49:507
11. Enokson B (1971) Chem Scripta 1:221
12. Emberger A (1975) Dissertation, Clausthal
13. Borchard W, Emberger A, Schwarz J (1978) Die Angew Makromol Chem 66:43
14. Svedberg T, Pedersen KO (1940) Die Ultrazentrifuge. Steinkopff-Verlag, Dresden
15. Flory PJ (1953) Principles of Polymer Chemistry. Cornell Univ Press, Ithaca, NY
16. Bloomfield VA (1976) Biopolymers 15:1243
17. Haase R (1963) Thermodynamik der irreversiblen Prozesse. Steinkopff-Verlag, Dresden
18. Callen HB (1962) Thermodynamics. John Wiley & Sons Inc., New York-London
19. Fujita H (1975) Foundations of Ultracentrifugal Analysis. John Wiley & Sons, New York-London
20. Cölfen H (1991) Diplomarbeit, Duisburg
21. Cölfen H, Holtus G, Borchard W, this issue II/III
22. Johnson P (1964) Proc Royal Soc A 278:527
23. Johnson P, Metcalfe JC (1967) Europ Polym J 3:423
24. Johnson P, King RW (1968) J Photograph Sci 16:82
25. Johnson P (1971) J Photograph Sci 19:49
26. Johnson P (1970) Photographic Gelatin I, editors Cox RJ Academic Press, London p 13
27. Holtus G (1990) Doktorarbeit, Duisburg
28. Holtus G, Borchard W (1989) Colloid Polym Sci 267:1133
29. Borchard W, Holtus G (1989) Colloid Polym Sci 267:1127
30. Haase R (1956) Thermodynamik der Mischphasen. Springer-Verlag, Berlin-Göttingen-Heidelberg
31. Borchard W (1982) presented at the DECHEMA meeting in Frankfurt
32. Unbehend M (1961) Dissertation, Aachen
33. Rehage G (1963) Kunststoffe 53:605
34. Borchard W (1978) Europ Polym J 14:661
35. Dusek K, Prins W (1969) Adv Polymer Sci 6:1

Authors' address:

Prof. Dr. W. Borchard
Angewandte Physikalische Chemie
Universität-GH-Duisburg
Lotharstr. 1
4100 Duisburg, FRG

Progress in Colloid & Polymer Science   Progr Colloid Polym Sci 86:92—101 (1991)

Swelling pressure equilibrium of swollen crosslinked systems in an external field. II:
The determination of molecular parameters of gelatin/water gels from the swelling pressure-concentration curves

G. Holtus, H. Cölfen, and W. Borchard

Angewandte Physikalische Chemie der Universität-GH-Duisburg, FRG

Abstract: Physically crosslinked watery gels of a dialyzed pigskin gelatin of type A are investigated by means of equilibrium runs at 10 and 20°C. From the data of the Schlieren patterns the swelling pressure-concentration curves can be calculated. It can be shown that equilibria are achieved in all cases and that the swelling pressure which is calculated for a given concentration is independent of the rotational speed selected for the experiment if all other conditions are kept constant. The results are described by the swelling equation of the Flory-Huggins type where the interaction paramter χ is allowed to depend linearly on the polymer concentration. Thus, two interaction constants and the network parameter have been calculated for each network by means of a nonlinear numerical iteration due to the Gauss-Jordan procedure. With these constants the swelling pressure-concentration curves are well described. The χ-parameter at low initial concentration of the polymer is very close to the value which has been experimentally determined in polymer solutions in the highly diluted range above the coil-helix transition range. Both χ-values reveal the influence of a highly branched structure of the polymer network. The calculated elastic modulus has nearly the same order of magnitude as that which has been obtained from the experimentally determined real part of the complex shear modulus of the same gel.

Key words: Thermoreversible gelation; physical network; swelling pressure; ultracentrifugation; Flory-Huggins equation

Introduction

The analytical ultracentrifuge is a very useful device to determine swelling pressures of gels, because it is possible to obtain several swelling pressure-concentration curves which are measured exactly under the same conditions in only one equilibrium run if a multiplace rotor is used. As the measuring method is very sensitive, thermodynamic properties of the gels can be calculated properly.

All the classical methods to determine swelling pressures of gels have the great disadvantage that only a single swelling pressure depending on the concentration can be obtained per measurement. Thus, it is very difficult to keep all the measuring conditions constant in all experiments which are needed to construct a swelling pressure-concentration curve. Some swelling pressure devices are described in the following text to demonstrate the principles.

Posnjak was the first to develop an apparatus to measure swelling pressures [1]. He allowed gelatin and rubber gels to swell under defined external swelling pressures and determined the gel volume in the heterogeneous equilibrium. Lloyd and Moran measured the pressure which was needed to remove a quantity of solvent from the gel [2]. The results of both described methods differed greatly in the equilibrium case, although they should have been the same. Prins and Pennings described an instrument by which the swelling pressure of the gel was transferred to a liquid column with a metal or polyethylene foil [3]. This device was improved by van de Kraats [4]. Borchard constructed an apparatus in which the sample was in contact with the solvent through a plate of sintered metal. The sample was screened from the pressure-transducing mercury by a thin foil. After a pressure was exerted on the mercury by means of a gas, the change in the volume of the gel was measured as a displacement of a column of mercury in a capillary [5]. Finally, a device was set forth where the swelling pressures of gels could be measured under isotropic conditions at various temperatures [6].

Swelling pressure measurements have been performed using an analytical ultracentrifuge as pressure generator (Borchard). A polystyrene/cyclohexane gel was compressed by means of a porous disc. The concentration could be calculated from the shift of this plate [7].

Although gels become increasingly important in industrial applications, only a few ultracentrifugal studies are known. The first investigations of gels were carried out by Svedberg [8]. Johnson made sedimentation velocity experiments with buffered agar and gelatin gels to investigate the influence of parameters like pH, ionic strength, initial concentration, and time of crosslinking on the sedimentation coefficient [9—11]. Further sedimentation equilibrium experiments with agar gels carried out by Johnson showed that the deformation of the swollen elastic system leads to a real equilibrium [12, 13]. Investigations of watery gels of DNA with a high molar mass by means of equilibrium runs were reported by Richard, who studied their properties under defined conditions [14—16]. Nearly spherical polybutadiene and polychloropropene gels were investigated by Lange to determine the degree of swelling and to calculate the average degree of polymerization of the chains between two connections of the network with that value [17]. By means of equilibrium runs Holtus studied watery gelatin gels [18, 19]. He found out that these gels form inhomogeneous networks below certain initial concentrations and that the gelation time of the samples has an influence on the swelling pressure curves. If the gelation time was taken long enough, no soluble parts of the dialyzed gelatin could be detected in the gels. The latest experiments with gels were carried out by Cölfen concerning the systems gelatin/water and κ-carrageenan/water [20, 21]. Both polymers are non dialyzed samples and are not completely incorporated into the polymer network of the gel after a long time of gelation. Thus, the gels consist of a physical polymer network which is embedded in a solution of soluble polymer molecules. These so-called soluble parts which remain in the state of the solution (sol) have an influence on the swelling pressure equilibria of the gels.

Method

A Beckman model E analytical ultracentrifuge equipped with Schlieren optics and a multiplexer for a six-place rotor was used. The Schlieren patterns are transmitted from a focusing screen in the film plane to a computer monitor by a video camera. The refractive index/distance curves can be digitized by means of a graphic tablet. The digitized traces are saved on a floppy disk [19].

If a gel is brought into a centrifugal field, a pressure gradient inside the network in the direction of the centrifugal force will be generated. The locally dependent pressure causes a concentration gradient in the gel due to the pressure gradient. Therefore, the meniscus gel/solvent has to shift in direction of the cell bottom as soon as the maximum solvent concentration due to the swelling equilibrium has established at the boundary gel/solvent [22]. It has been shown that the shift leads to real equilibrium states [12, 18—21]. Because of the high optical density of the samples at higher polymer concentrations the concentration gradient cannot be observed in the whole gel phase. Therefore, an approximation has to be used to calculate the concentration in dependence on the radial distance from the axis of rotation r [8, 19—21]. In the equilibrium deformation of an isotropic gel the swelling pressure for every polymer concentration can be calculated with the following equation, assuming additivity of the volumes of the pure components [22]:

$$\Pi_s = \omega^2 \int_{r_m^{g/s}}^{r_i} r\rho(r) \left[1 - \frac{1}{\rho(r)\tilde{V}_{01}} \right] dr , \qquad (1)$$

with Π_s as swelling pressure, ω is the angular frequency, $\tilde{V}_{01}$ is the partial specific volume of the pure solvent, $\rho(r)$ the local density of the gel, r the radial distance from the axis of rotation to any point (index m stands for the meniscus, g/s for gel-solvent, i characterizes any distance from the axis of rotation to any point i inside the gel).

The statistical theory of swelling of a nonelectrolyte polymer is related to the theory of polymer solutions and the theory of rubber elasticity. If we start from a slightly modified Flory-Huggins equation, the difference of the chemical potentials $\Delta\mu_1$ of the solvent inside the binary gel μ_1 and that of the pure solvent μ_{01} is given by [23—31]

$$\frac{\Delta\mu_1}{RT} = \ln(1 - w_2) + w_2 + \chi w_2^2$$

$$+ \frac{1}{z_w} \cdot [A\eta w_2^{1/3} - B w_2] , \qquad (2)$$

with the quantities: R being the universal gas constant, w_2 the weight fraction of the polymer, T the thermodynamic temperature, χ the interaction parameter, z_w the ratio between the number average molar mass of a chain segment between two junction points of the network, $\overline{M}_c$, and the molar mass of the solvent, M_1; and η is the ratio of the mean value of the chain end-to-end distance of the chain in the network and that of the free chain; A, B are characteristic network constants according to the different network theories.

If the crosslinking of the network has been performed in solution, the degree of swelling of this state has to be introduced, indicating that in the reference state of the dry polymer the network chains are in a somewhat supercoiled state. Therefore, η is given for the gelation approximately by $\eta = (w_2^0)^{2/3}$, where w_2^0 is the weight fraction of the polymer in the original polymer solution. A possibly occurring change of the end-to-end distance of the polymer chain during crosslinking has been ignored [27].

In the theory it is assumed that, in calculating the mixing entropy, the gel can be considered as a polymer chain with an infinitely high molar mass. The network chains have to be fixed to the network junctions permanently. If the physical network of the thermoreversible gel is formed by entanglements, these also have to be permanent during the time of attainment of the heterogeneous/continuous equilibrium.

In the original form of Eq. (2) the volume fraction is the variable for the concentration which has been exchanged by the weight fraction in this treatment, as was done also by Scholte for other systems [32]. This is a further approximation required by unknown parameters of the dry, glassy polymer

which are necessary to calculate the volume fractions which depend on temperature. Therefore, all deviations entering from this substitution influence the network-characterizing parameters. The most important approximation is probably that we treat the amphoteric polyelectrolyte copolymer as a non electrolyte homopolymer which is soluble and swellable in water. In our treatment, we neglect term B, which has been much debated [27, 33] and which does not give an essential improvement in the description of our results.

The swelling pressure can be considered as an osmotic pressure and is related to the difference of the chemical potentials of the solvent by [36]:

$$\Delta\tilde{\mu}_1 = -\Pi_s \tilde{V}_1 = -\Pi_s M_1 V_1 = M_1 \Delta\mu_1 , \qquad (3)$$

where Π_s is the swelling pressure. $\tilde{V}_1$ and V_1 are the partial specific and partial molar volumes of the solvent in the gel. $\Delta\tilde{\mu}_1$ is the difference between the specific chemical potential of the solvent in the gel and the specific chemical potential of the pure solvent, M_1 is the molar mass of the solvent.

Under the assumptions mentioned, the combination of Eqs. (2) and (3) leads to Eq. (4), which allows to calculate swelling pressure curves in dependence of the mass fraction of the polymer:

$$-\frac{\Pi_s V_1}{RT} = \ln(1 - w_2) + w_2 + \chi w_2^2$$

$$+ \frac{1}{z_w} [A (w_2^0)^{2/3} w_2^{1/3}] . \qquad (4)$$

Only in few systems has the χ-parameter been found to be a constant in rather limited ranges of concentration [27, 34, 35].

If we consider the quantity χ as depending on concentration, we have in the next simplest case [34, 36, 37]:

$$\chi = \chi_0 + \chi_1 w_2 . \qquad (5)$$

Calculations for the system gelatin/water have shown that higher terms in the dependence of χ upon concentration are of no significant influence.

Substituting Eq. (5), gathering all network characterizing parameters in a single term C, and replacing the partial molar volume of the solvent V_1 by the molar volume V_{01}, Eq. (6) is obtained, which is used for the nonlinear numerical iteration due to the Gauss-Jordan procedure. We only have to know the experimentally determined swelling pressure-concentration curve to calculate the constants.

$$-\frac{\Pi_s V_{01}}{RT} = \ln(1 - w_2) + w_2 + \chi_0 w_2^2$$

$$+ \chi_1 w_2^3 + C w_2^{1/3} . \qquad (6)$$

With the relations mentioned it is possible to calculate the mean molar mass of the network chains $\overline{M}_c$ from C if A and η can be estimated. The Young modulus E is related to the quantity C by

$$E = \frac{RTC w_2^{1/3}}{V_{01}} , \qquad (7)$$

where all quantities are already known. As for rubber elastic materials, the Poisson ratio is nearly 0.5, and the shear modulus G is given by $G \approx E/3$.

Experimental

We investigated a dialyzed photographic gelatin Type A (which is a polypeptide derived from pigskin collagen by means of an acid treatment). The gelatin has a water content of 13.8% by wt. and a broad distribution of molar masses, so that it can be considered as a copolymer. Because of the dialyzation no soluble parts occur which influence the swelling pressure equilibria. The investigated samples are watery gels in the concentration range of 1.9 up to 9.8% by wt.

The gelatin is dissolved in distilled water. To prevent growth of bacteria 0.15 ml of a 5% by wt. methanol solution of chlorophenol is added per gram gelatin granules. The granules are allowed to swell over night in a refrigerator at about 7°C. The gelatin is heated to 50—60°C for 45—60 min to prepare a sample for an experiment. The higher the gel concentration, the longer the time needed. During the heating the bottle is shaken a few times to accelerate the homogenization of the sol. Before mounting the cells, their walls and windows are prepared with a thin film of silicon oil to avoid the adhesion of the gelatin gel. The ultracentrifugal cells are filled with about 0.4 ml of the hot solution while the sector of the centerpiece is in such a position that the gravitational field acts in the same

direction as the later centrifugal field. After filling of the cells they are cooled to the desired gelation temperature while the position of the centerpieces is kept constant.

The sol is allowed to gel for 48 h. Then the cells are placed into the rotor, which is already conditioned at the same temperature as that in the rotor chamber. After a renewed conditioning of the rotor with the cells for 24 h the equilibrium run is started, accelerating the rotor to the selected rotational speed. Schlieren patterns are drawn after various time intervals. When no further movement of the meniscus gel/solvent is observed the equilibrium is considered as having been reached. Then the rotational speed is raised by 4000 rotations per minute (RPM) until the meniscus gel/solvent has clearly moved in the direction of the bottom of the cell; afterwards the speed is lowered to the former selected speed. The run is stopped after the new equilibrium is reached. Thereby the equilibria are reached from two different sides.

Because the maximum degree of swelling of the various gels at the desired temperatures has to be known for the calculation of the swelling pressure-concentration curves defined masses of the gels are allowed to swell in a large amount of distilled water in a glass vessel at the same temperature as that of the measurement in the ultracentrifuge. To establish a great surface of the gel a lot of small gel pieces are cut out of the gel. For every gel concentration 12 samples are prepared to allow one measurement per day over a period of 12 days. To determine the mass of the swollen sample the content of the bottle is poured on a sieve of stainless steel which is sufficiently fine to contain the gel pieces on its surface. The amount of adhering water is absorbed by tissues placed below the sieve. With the known polymer content of the gel the weight fraction of the swollen sample can be calculated. Because, after long periods of time, the gels are soluble in a large quantity of pure water, the experiments are continued until the gel is obviously dissolved.

Results and discussion

If the ratio of the mass fraction of the polymer in the gels $w_2(t = 0)$ at the beginning of the swelling experiment and the corresponding value $w_2(t)$ after a swelling time t is plotted versus t, as shown in Fig. 1, a plateau value is reached. This plateau ratio

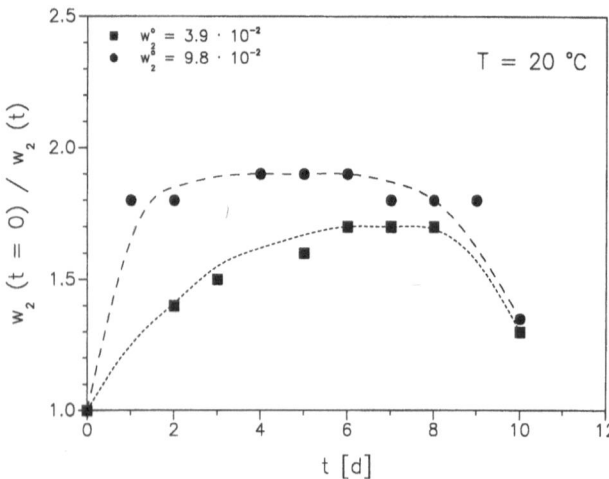

Fig. 1. Swelling of different concentrated gelatin/water gels in large amounts of water at 20°C in dependence on time t (see text)

is expected if the swelling equilibrium is established. But after some days a dissolution of the gel in a large amount of water is observed. Therefore, the mass of the gel from which $w_2(t)$ is calculated will be decreased, which leads to apparently lower values of the ratio $w_2(t = 0)/w_2(t)$.

In order to keep the experimental error for the determination of the stationary degree of swelling of the samples as small as possible the value for the maximum degree of swelling is taken as soon as the maximum value for $w_2(t = 0)/w_2(t)$ is reached.

The sedimentation of all investigated gels corresponds to a real equilibrium deformation. This could be proven by the result that it is possible to reach the equilibrium position of the meniscus gel/solvent from a lower and a higher rotational speed compared with the selected one [18, 19]. This proof is due to the principle that it is possible to reach any heterogeneous or continuous equilibrium from different sides if the set of variables describing the system is not changed. This principle of path independence has to be fulfilled for true equilibria.

The swelling pressure which can be calculated for a certain concentration from Eq. (1) has to be independent of the angular velocity ω, i.e., the rotational speed if real equilibria are established. This means that the swelling pressure-concentration curves must coincide, in the equilibrium case at different rotational speeds if all other experimental parameters like temperature, concentration, and

prehistory of the gel are kept constant. The only difference is the concentration range covered which corresponds directly to the maximum swelling pressure that is present at the bottom of the cell and which is, of course, dependent on the rotational speed. In Fig. 2 this is demonstrated for the case of gelatin/water gels which have been investigated at different rotational speeds and two runs at a constant speed.

If the parameters χ_0, χ_1 and C are calculated from the experimental swelling pressure curves by means of Eq. (6), it is of interest if the experimentally determined swelling pressure-concentration curves are well described by the calculated ones. Only a good agreement of the experimental and the calculated curves allows us to use the constants χ_0, χ_1, and C for further determination of molecular parameters. Figure 3 shows the experimental points and the calculated swelling pressure-concentration curves for a gelatin/water gel with different initial concentrations. All intersecting curves pass the experimental points in the range of experimental accuracy, which is ± 0.006 bar concerning the swelling pressure. This shows that the iteration due to the Gauss-Jordan procedure starting from Eq. (6) has led to reasonable results for the parameters characterizing the gels. The values are presented in Table 1.

It can be seen by comparison of the χ_0- and χ_1-values that, with increasing initial concentration of gelatin, the concentration-independent term χ_0 is slightly decreasing, whereas χ_1 and C are increasing for both temperatures. It is expected that the branching of polymer chains and the crosslinking density and, thereby, its term C increases when the number of the network chains with possible contacts per chain is enlarged. This increase in branching and crosslinking density seems also to slightly influence the concentration dependence of the interaction parameter. This result is in agreement with the findings of chemically crosslinked gels of the system PS-cyclohexane where a decrease of χ_0 and an increase of χ_1 with the increase of the content of the crosslinking agent has been found [26].

Similar behavior is reported for the system cyclohexane-branched PS (star and comb model polymers) where the concentration dependence of χ is influenced by the degree of branching [38]. It is interesting to compare the χ_0- and χ_1-values with those of Candau et al., because the magnitude of their χ_0- and χ_1-values is roughly the same at 30°C

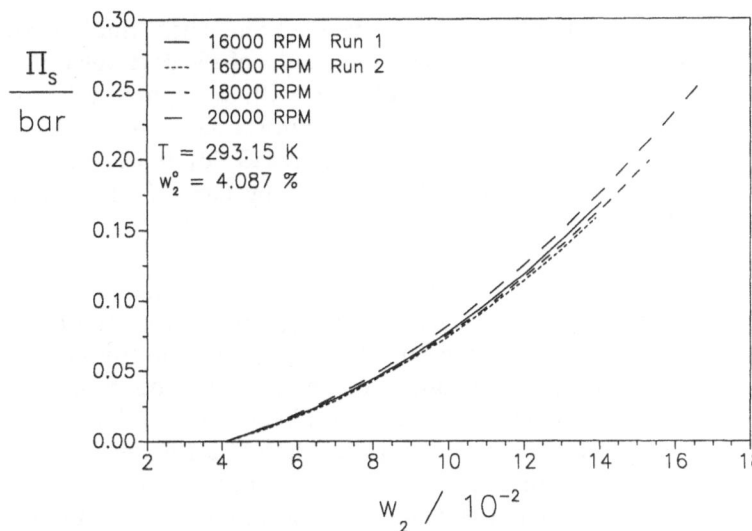

Fig. 2. Swelling pressure Π_s of the system gelatin/water vs concentration of gelatin for different rotational speeds. w_2 is the weight fraction of the polymer, w_2^0 its initial value; $T = 293.15$ K, which is the temperature of gelation and of the experiment; RPM = rotations per minute (see text)

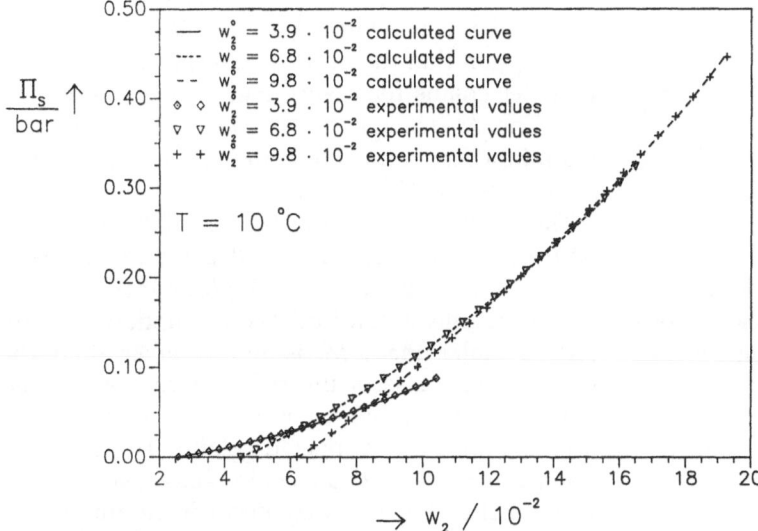

Fig. 3. Comparison of experimentally determined swelling pressures Π_s of the system gelatin/water vs concentrations of gelatin with calculated ones. w_2 is the weight fraction of the polymer; w_2^0 its initial value. $T = 283.15$ K, which is the temperature of gelation and measurement

under conditions near the θ-temperature of the system cyclohexane/linear PS. Thus, we register a structural effect in the $\chi(w_2)$-dependence which may even be more pronounced, because in the system gelatin/water, not only branching, but also crosslinking takes place. The linear extrapolation of χ_0 and χ_1 in Table 1 vs the concentration w_2^0 to the value $w_2^0 = 0$ leads to $\chi_0(w_2^0 = 0) = 0.4975$ and $\chi_1(w_2^0 = 0) = 0.362$. These interaction parameters correspond to unbranched and uncrosslinked structures which are difficult to measure directly because of

the aggregation of the polymer [39]. But it is found that both values do not depend on the temperature between 283.15 and 293.15 K.

With the χ_0- and χ_1-values the χ-parameters have been calculated by use of Eq. (5) for the initial weight fractions of the polymer $w_2 = w_2^0$ which are presented in Table 2.

Although we have found a slightly different concentration dependence of the interaction parameters due to the initial concentration of the polymer the actual quantities for the different

Table 1. Network parameters χ_0, χ_1, and C for different temperatures T and different initial weight fractions of the polymer w_2^0 for the system gelatin/water (see text)

$w_2^0 \cdot 10^2$	1.96	3.92	6.87	7.81	9.81	
χ_0	0.492	0.487	0.477	——	0.470	$T = 283.15$ K
χ_1	0.387	0.412	0.451	——	0.481	$T = 283.15$ K
$C \cdot 10^4$	0.059	0.228	1.030	——	2.146	$T = 283.15$ K
$_0$	0.493	0.489	0.481	0.480	——	$T = 293.15$ K
χ_1	0.386	0.412	0.448	0.453	——	$T = 293.15$ K
$C \cdot 10^4$	0.028	0.150	0.492	0.595	——	$T = 293.15$ K

Table 2. Concentration dependent χ-parameter for the system gelatin/water for different initial weight fractions of the polymer w_2^0 and temperatures T calculated with Eq. (5) (see text)

$w_2^0 \cdot 10^2$	1.96	3.92	6.87	7.81	9.81	
χ	0.497	0.498	0.497	——	0.500	$T = 283.15$ K
χ	0.497	0.498	0.497	0.497	——	$T = 293.15$ K

w_2^0-values are nearly the same and are independent of the temperature. These values are mean values for the system copolymer/solvent. The segments of the native polypeptide differ greatly in their structures and charges [40]. There are segments for which water is a good solvent ($\chi < 0.5$) and segments for which water is a poor solvent ($\chi > 0.5$). The system seems to be, on average, close to a miscibility gap where polymers have a tendency to associate and to gel [41, 42].

From osmotic measurements of low concentrated gelatin solutions in the temperature range 25—70 °C, a temperature independent value $\chi = 0.483 \pm 0.024$ has been found [43, 44]. In the error limit this value fits well with the values listed in Table 2 although the extrapolated value $\chi_0(w_2^0 = 0)$ is a little bit higher, perhaps due to the different temperature regions above and below the temperature for the coil-helix transition.

Furthermore, it is seen that the network parameters C at the temperature 283.15 K are nearly twice as large as those at 293.15 K. This is explained by a higher crosslinking density at lower temperatures at

constant polymer concentration. These findings are in agreement with the results of a watery gelatin gel of 4% by wt. polymer with a slightly different prehistory where the complex shear modulus G^* has been measured directly [45, 46]. In this case the real part of G^* was larger by factor 3 at 283.15 K with respect to that at 293.15 K for a frequency of 1 Hz.

For gels in the rubbery state an increase of the equilibrium shear or Young's modulus proportional to T following Eq. (7) is expected if the crosslinking density remains constant. From the frequency dependence of G^* it has been derived that gels of gelatin/water are in the state of a rubber [47—49].

If the swelling pressure—concentration curves are evaluated as before, but assuming a concentration independent χ-parameter, negative z_w-values are obtained which are physically meaningless [19]. This was also pointed out by Emberger [50].

In a rough estimation from the C-values of Table 1, the mean molar mass between the junction points of the network $\overline{M}_c$ is calculated under the assumption that the network consists of endlinked polymer chains for a very high functionality ($f \rightarrow \infty$). The $\overline{M}_c$-average molar masses between network junction points are gathered in Table 3. $\overline{M}_c$ is related to z_w by $\overline{M}_c = z_w M_1$.

The number average molar mass of the molecules occurring in the denatured collagen named gelatin is between $5 \cdot 10^4$ and $5 \cdot 10^5$ g/mol [51]. We state that, especially at low initial concentrations, the apparent molar mass $\overline{M}_c$ is much larger than the mean molar mass of the polymer chains the network is built of, which is 68000 g/mol (number average) for the investigated gelatin. This is a contradiction if all polymer chains have become network chains. But this may occur if the number of the effective network chains is smaller than the total number of the chains before crosslinking. For concentrations below 4% by wt. the relatively stiff polymer chains have to aggregate to build up a long chain between the network junctions. This is only possible if many single chains are partly overlapping along each other, making a highly branched structure where only the longest chain connects the junction points. The kind of the junctions and their number along each chain are not yet known definitely [49].

As a polymer solution in the dilute region is not homogeneous with respect to the local density of the polymer, we have to assume that this inhomogeneity exists also in the case of aggregating

Table 3. Number average molar mass of a chain between two junction points of the network $\overline{M}_c$ for different initial concentrations w_2^0 of gelatin/water gels (see text)

			$f \to \infty$			
$w_2^0 \cdot 10^2$	1.96	3.92	6.87	7.81	9.81	
$\overline{M}_c$ [g/mol]	217232	90792	29114	——	17829	$T = 283.15$ K
$\overline{M}_c$ [g/mol]	457738	138004	60951	55227	——	$T = 293.15$ K
			$f = 3$			
$w_2^0 \cdot 10^2$	1.96	3.92	6.87	7.81	9.81	
$\overline{M}_c$ [g/mol]	72411	30264	9705	——	5943	$T = 283.15$ K
$\overline{M}_c$ [g/mol]	152579	46001	20317	18409	——	$T = 293.15$ K

polymers. Therefore, in this concentration range highly branched and also crosslinked regions are interlinked by a few very long molecules. At higher polymer concentrations $\overline{M}_c$ is lower than the mean molar mass of the chains. In this case the coiled polymer chains overlap in the states of a sol and gel. Therefore, there is much more of a chance that the branching of the molecules leads to a homogeneous local polymer density.

If $\overline{M}_c$ is calculated under the assumption of fluctuating network junctions with $f = 3$ (which is listed in Table 3), we state that only for the lowest initial concentration is the value of $\overline{M}_c$ larger than the average molar mass of the polymer. This means that, even in this case, longer chains have to be produced by aggregation of the polymer chains. We know that most of the polymer during the isothermal gelation at low polymer concentrations first undergoes a rapid coil-helix transition with subsequent enlargement of the molecules by aggregation. At a later stage of the aggregation a polymer network is formed which can be characterized by its elastic properties [47—49].

It is now possible to calculate Young's modulus from the quantities C in Table 1 by means of Eq. (7) and the approximation that the Poisson ratio for rubbery materials is nearly 0.5. In this case, Young's modulus E is related to the shear modulus by $G \approx E/3$. The equilibrium shear moduli G can now be compared with the experimentally determined storage shear modulus G' which is the real part of the complex shear modulus. All values as far as available are listed in Table 4 for different temperatures and initial concentrations. It is conceivable that the G'-values at 1 Hz are larger by a

factor 10 than the G-values from the sedimentation-diffusion equilibrium of the gels. This can easily be explained by the frequency dependence of the viscoelastic property G^*.

It can be derived by means of a creep experiment that G' decreses by factor 2 after 24 h. Therefore, it is not astonishing that the equilibrium modulus G which has been determined in the swelling pressure experiment lasting several days is much smaller.

Finally, we have to ask if it is allowed to apply the swelling theory of Flory to a highly branched network structure. If we know that only some of the network chains are elastically active, then the contribution of the network term to Eq. (6) cannot be calculated from the number of the "primary chains" present in the network under the assumption that every chain becomes a network chain during gelation. Thus, the total mass of the polymer m_2 consists of the mass of the network $m_{2,\text{net}}$ and that of the branched but not elastically effective material attached to the network $m_{2,\text{branch}}$. From the relation $m_{2,\text{net}} = (\nu_e \overline{M}_c)_{\text{net}}$, where ν_e is the effective number of network chains in moles we do not know $m_{2,\text{net}}$, because only ν_e or $\overline{M}_c$ is known, but the inequality $m_2 > m_{2,\text{net}}$ tells us that $(\nu_e \overline{M}_c)_{\text{net}}$ is smaller than $(\nu_e \overline{M}_c)$ calculated from the endlinking of all chains. Therefore, it is not possible to decide between the model of very few, very long chains or a few long chains which are perhaps nearly as long as the primary chains. If the number of the junction points is very low there will be an inhomogeneity of the crosslinking density.

For a further discussion model calculations of swelling pressure curves have to be made which

Table 4. Elastic mdoulus E and shear modulus G for gelatin/water gels with different initial mass fractions w_2^0 at 293.15 and 283.15 K. Calculated and experimental values[a]). Equilibrium values taken from [49][b]). This concentration is measured at 285.15 K after a gelling time of 22 d*) (see text)

$w_2^0 \cdot 10^2$	E [Pa] calc.	G [Pa] calc.	$T = 293.15$ K E' [Pa] exp.	G' [Pa] exp.	$w_2^0 \cdot 10^2$
1.96	$0.0101 \cdot 10^4$	$0.0034 \cdot 10^4$	$0.1008 \cdot 10^4$	$0.0336 \cdot 10^4$	2.00[a])
3.92	$0.0689 \cdot 10^4$	$0.0230 \cdot 10^4$	$0.4020 \cdot 10^4$	$0.1340 \cdot 10^4$	4.25[a])
6.87	$0.2719 \cdot 10^4$	$0.0906 \cdot 10^4$			
7.81	$0.3442 \cdot 10^4$	$0.1147 \cdot 10^4$			
$w_2^0 \cdot 10^2$	E [Pa] calc.	G [Pa] calc.	$T = 283.15$ K E' [Pa] exp.	G' [Pa] exp.	$w_2^0 \cdot 10^2$
1.96	$0.0206 \cdot 10^4$	$0.0069 \cdot 10^4$			
3.92	$0.1011 \cdot 10^4$	$0.0337 \cdot 10^4$			
6.87	$0.5498 \cdot 10^4$	$0.1833 \cdot 10^4$	$1.4823 \cdot 10^4$	$0.4941 \cdot 10^4$	6.94[b])
9.81	$1.2939 \cdot 10^4$	$0.4313 \cdot 10^4$			

*) We thank Dipl. Chem. A. Michalczyk for the measurement of the real part of the complex shear modulus[b]).

will perhaps lead to an understanding of intersecting swelling pressure-curves stemming of gels with different initial concentration. The application of the Flory theory seems to be reasonable as long as the local density of the polymer is the same in all volume elements; this is clearly fulfilled beyond the critical concentration of the coil overlap. For the lower polymer concentrations <6% by wt. gelatin, this may be questioned.

Conclusion

The deformation of a gel in a centrifugal field yields an experimentally determined swelling pressure-concentration curve which starts at the swelling equilibrium. The application of a slightly modified Flory-Huggins equation with an interaction parameter χ which depends linearly on the polymer concentration leads to reasonable results for χ and the network term C for the system gelatin/water. The concentration dependence of χ reveals structural influence caused by branching and crosslinking. The χ-value which has been extrapolated to zero initial concentration is very close to the value which is derived from osmotic measurements. The Young modulus calculated from the network term is in agreement with the experimentally determined one. Therefore, we sup-

pose that this new procedure to determine thermodynamic properties of gels can be applied to other systems with success. Furthermore, an explanation is given of why the apparent number average molar mass of the chains between two junction points is found to be much higher than the value for the primary chain for gels with low initial concentrations. The inhomogeneities in gelatin networks below certain initial concentrations, which have been reported before, are probably due to an inhomogeneously crosslinked and highly branched structure [18, 19].

Acknowledgements

The authors are very grateful to the Deutsche Forschungsgemeinschaft (DFG) and to the Max-Buchner Forschungsstiftung for financial support.

References

1. Posnjak E (1912) Koll Chem Beiheft 3:417
2. Lloyd DJ, Moran T (1934) Proc Roy Soc 147A:382
3. Pennings AJ, Prins WJ (1961) Polymer Science 49
4. van de Kraats EJ (1968) Rec Trav Chim Pays Bas 87:1137
5. Borchard W (1966) Dissertation, Aachen
6. Borchard W, Emberger A, Schwarz J (1978) Die Angewandte Makromolekulare Chemie 66, Nr. 986:43

7. Borchard W (1975) Progr Colloid Polym Sci 57:39
8. Svedberg T, Pedersen KO (1940) Die Ultrazentrifuge, Steinkopff, Dresden
9. Johnson P (1964) Proc Royal Soc A 278:527
10. Johnson P, Metcalfe JC (1967) Europ Polym J 3:423
11. Johnson P, King RW (1968) J Photograph Sci 16:82
12. Johnson P (1971) J Photograph Sci 19:49
13. Johnson P (1970) Velocity and equilibrium aspects of the sedimentation of agar gels in Photographic Gelatin I, Academic Press, London:13
14. Richard AJ (1983) Biopolymers 22(3):935
15. Richard AJ (1984) Biopolymers 23(7):1307
16. Richard AJ, Westkaemper RB (1986) Biopolymers 25(10):2017
17. Lange H (1986) Colloid Polym Sci 264:488
18. Holtus G, Borchard W (1989) Colloid Polym Sci 267:1333
19. Holtus G (1990) Dissertation, Duisburg
20. Cölfen H (1991) Diplomarbeit, Duisburg
21. Cölfen H, Borchard W this issue
22. Borchard W this issue
23. Flory PJ (1942) J Chem Phys 10:51
24. Huggins ML (1943) Ann NY Acad Sci 44:431
25. Staverman AJ (1962) Thermodynamics of Polymers in Flügge S (1962) Encyclopedia of Physics/Handbuch der Physik, Springer Berlin-Göttingen-Heidelberg 1962
26. Borchard W (1975) Habilitation, Clausthal
27. Dusek K, Prins W (1968) Adv Polym Sci 6:58
28. Candau S, Bastide J, Delsanti M (1982) Structural Elastic and Dynamic Properties of Swollen Polymer Networks in Dusek K (1982) Polymer Networks, Springer Berlin-Heidelberg-New York 27
29. Graessley WW (1975) Macromolecules 8:186, 865
30. Edwards SF (1971) The Statistical Mechanics of Rubbers in Chömpff & Newman (1971) Polymer Network Structure and Mechanical Properties, Plenum Press, New York-London
31. Kilian HG, Schenk H, Wolff S (1987) Colloid Polym Sci 265 Nr. 5:410
32. Scholte TG (1971) J Polym Sci 9A2:1553
33. Staverman AJ (1982) Properties of Phantom Networks and Real Networks in Dusek K (1982) Polymer Networks, Springer Berlin-Heidelberg-New York 1982:73
34. Rehage G (1964) Kolloid-Z u Z Polymere 194:16 and 196:57
35. Koningsveld R, Kleintjens LA (1971) Macromolecules 4:5, 637
36. Haase R (1956) Thermodynamik der Mischphasen, Springer Berlin-Göttingen-Heidelberg
37. Adames W, Michalczyk A, Borchard W (1989) Europ Polym J 25 Nr. 9:951
38. Candau F, Strazielle C, Benoit H (1976) Europ Polym J 12:95
39. Boedtker H, Doty P (1954) J Phys Chem 58:968
40. Fietzek PP, Kühn K (1975) Cellular Biochem 8:141
41. ter Meer HU (1985) Thermoreversible Gelierung: Carrageenan; Agarose; Alginate and Pektin in Burchard W (1985) Polysaccharide, Springer
42. Stauffer D, Coniglio A, Adam M (1982) Gelation and Critical Phenomena in Dusek K (1982) Polymer Networks, Springer Berlin-Heidelberg-New York 103
43. Keese AS (1978) Diplomarbeit, Clausthal
44. Borchard W, Keese A (1979) presented at the IUPAC Meeting Mainz
45. Borchard W, Bergmann K, Rehage G (1976) Investigations of Gelation Phenomena in Aqueous Gelatin Solutions in Photographic Gelatin II, Academic Press, London:57
46. Borchard W, Bergmann K, Emberger A, Rehage G (1976) Progr Colloid Polym Sci 60:20
47. Burg B (1988) Dissertation, Duisburg
48. Burg B, Borchard W (1988) Optical and Viscoelastic Properties of Gelatin-Water During Gelation in Lemstra PJ & Kleintjens LA (1988) Integration of Fundamental Polymer Science And Technology 3, Elsevier Applied Science London-New York 323
49. Borchard W, Burg B (1990) Progr Colloid Polym Sci 83:200
50. Emberger A (1975) Diplomarbeit, Clausthal
51. Rose PI (1987) in Encyclopedia of Polymer Science and Engineering, Volume 7, 2nd Edition, John Wiley & Sons:499

Authors' address:

Prof. Dr. W. Borchard
Universität-GH-Duisburg
Angewandte Physikalische Chemie
Lotharstr. 1
4100 Duisburg 1, FRG

Progress in Colloid & Polymer Science　　　　　Progr Colloid Polym Sci 86:102—110 (1991)

Swelling pressure equilibrium of swollen crosslinked systems in an external field. III:
Unsolved problems concerning the systems gelatin/water and κ-carrageenan/water

H. Cölfen and W. Borchard

Angewandte Physikalische Chemie der Universität-GH-Duisburg, FRG

Abstract: Different concentrated thermoreversible gelatin/water and κ-carrageenan/water gels prepared from non-dialyzed polymer material are set into an ultracentrifugal field and investigated by means of equilibrium runs at 20°C. The refractive index gradient and the movement of the meniscus gel/solvent during the process of swelling and deswelling is observed with the Schlieren optical system. From the Schlieren patterns at the sedimentation-diffusion equilibrium the swelling pressure can be calculated by use of an approximation based on the mass balance. Soluble parts of the polymer which are not incorporated into the network even after long gelation times could be detected by means of the Schlieren patterns in both systems. For the system gelatin/water a superposition of a sedimentation-diffusion equilibrium of the soluble parts and of the swelling pressure equilibrium of the gel could be detected. In this case the swelling pressure equilibrium is clearly influenced by the presence of soluble parts. In the case of κ-carraagee-nan/water a part of the sol associates outside the gel phase during the sedimentation and moves towards the gel surface.

Key words: Swelling pressure; ultracentrifugation; physical networks; soluble parts; thermoreversible gel

1. Introduction

Water sensitive networks like gelatin/water and κ-carrageenan/water have become more and more industrially important in recent years, for applications beside those of the food industry. Therefore, it is important to understand the process of gelling and the thermodynamic properties of these thermoreversible gels. One useful device to determine thermodynamic properties of gels is the analytical ultracentrifuge, because it makes it possible to obtain swelling pressures in a range of concentrations from only a single equilibrium run [4—8]. A theory for the deformation of a gel remaining in an isotropic state in the centrifugal field was developed in part I of our contribution in this issue. The theory was applied to watery gels of a dialyzed pigskin gelatin in part II. By means of a slightly modified Flory-Huggins equation, thermodynamic proper-

ties of the investigated gels could be obtained. But there remain other unsolved problems in the description of the behavior of gels in the ultracentrifugal field. Especially if the whole polymer material is not completely incorporated into the network, even after long gelation times, very complicated phenomena can occur. We want to present some of them in this article concerning the systems gelatin/water and κ-carrageenan/water.

2. Method

A Beckman Model *E* analytical ultracentrifuge equipped with Schlieren optics is used for the experiments described partly in contribution II in this issue [8]. A multiplexer for the use of a multicell rotor has been installed, and the optical system, the vacuum system, and the temperature control system have been modified [4, 5].

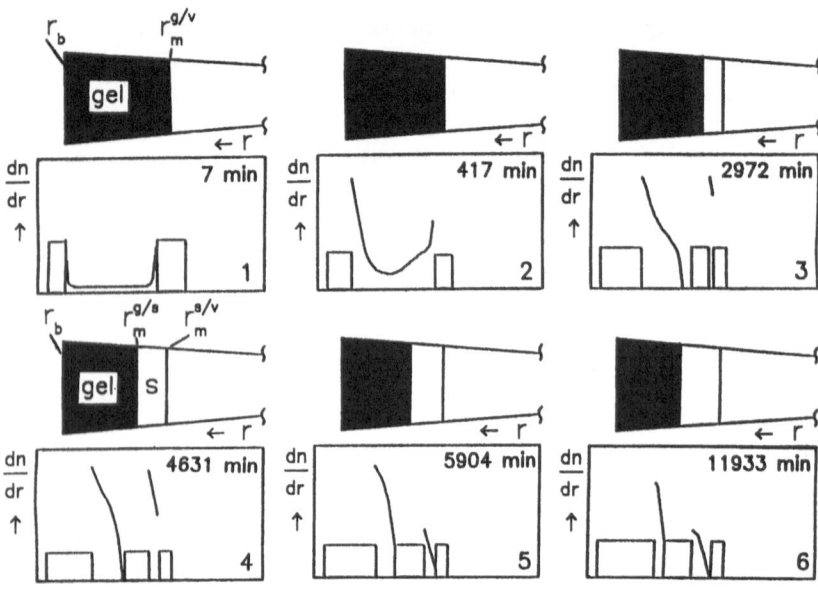

Fig. 1. Schlieren patterns and the corresponding schematic representation of the phases in a mono sector cell vs radial distance r during different stages of an equilibrium run of the system gelatin/water (see text). The phase boundaries are gel/vapor (g/v), solution(s)/vapor (s/v), and gel/solution (g/s). The index b stands for the bottom of the cell, m for the meniscus

Different concentrated gels of the investigated systems are analyzed in the ultracentrifugal field in a six-place rotor by means of equilibrium runs at 20 °C. With the Schlieren optical system the concentration gradient and the movement of the meniscus gel/solvent during the process of swelling and deswelling can be obtained. In Fig. 1 the Schlieren patterns, which are the curves where the gradient of the refractive index dn/dr is plotted vs the radial distance from the axis of rotation are shown. For different times during the process of deswelling of a gel different stages are reached which are schematically represented for a mono sector cell until equilibrium is attained.

The phase boundaries are shown as relatively broad stripes in the Schlieren patterns. They are not completely drawn into the Schlieren patterns in order to save disk space when the patterns are digitized. The broad meniscus is caused by the effect of interfacial tension. The adhesion of the gel at the cell walls can mainly be avoided by preparing the cell walls with PTFE spray, which establishes a thin film. In the case of a broad meniscus the middle of the meniscus is used for the calculations.

After the experiment is started only a single meniscus is present which is the interface between gel and vapor (Fig. 1.1). A few hours later a concentration gradient in the gel phase has appeared without a movement of the meniscus (Fig. 1.2). After approximately 1 day the deswelling of the gel

has started which can be observed by the occurrence of the moving meniscus gel/solvent (Fig. 1.3). In contrary to the dialyzed gelatin used before [7], it is stated that also a concentration gradient in the sol phase is present which is formed as soon as the meniscus of the gel starts to move. This gradient is caused by polymer material which is not built into the network, and is called the soluble part. The meniscus gel/solvent shifts in the direction of the cell bottom with increasing time until the equilibrium is reached (Fig. 1.4—1.6). Furthermore, it can be seen that the concentration gradient cannot be observed in the total gel phase, which has already been reported in the literature [1—3, 5, 6]. The broad range at the cell bottom which marks the area where the refractive index gradient of the polymer cannot be detected anymore becomes broader with increasing time. The range is caused by the high optical density of the samples at higher polymer concentrations.

In order to calculate the concentration of the network at every point in the cell an approximation for a binary system was used which is based on the cell geometry, the mass balance, and the assumption of a linear concentration gradient of the polymer in the gel phase. The linear concentration gradient in the equilibrium case has been demonstrated by Johnson [2, 3] and Holtus [5]. Because the investigated systems with the presence of soluble parts have at least to be treated as ternary ones, the

following approximation which has been used for binary gels can only be considered a rough one.

To calculate the concentration of the polymer as a function of the radial distance from the center of rotation the dependence of the specific volumes of the gels from the concentration w_2 at a given temperature has to be known; it was found to be linear in the considered concentration range for the investigated gels $\tilde{V} = aw_2 + b$, where a is the slope and b is the intercept. As the density of the gel ρ is given by $\rho = 1/\tilde{V}$, we obtain $\rho = (aw_2 + b)^{-1}$, which we call the density/concentration relation. Also, the initial concentration of the gel w_2^0, the density of the pure solvent ρ_{01}, and the radial distance from the axis of rotation $r_m^{g/s}$ of the meniscus gel/solution in the equilibrium case (which is available from the Schlieren pattern) are needed. With the known geometry of the sector-shaped cell with the sector angle φ and the height h, the volumes of the different phases can be calculated:

$$V_{gel}^0 = [(r_b)^2 - (r_m^{g/v})^2]h\pi\frac{\varphi}{360} , \tag{1}$$

and

$$V_{gel} = [(r_b)^2 - (r_m^{g/s})^2]h\pi\frac{\varphi}{360} , \text{ and} \tag{2}$$

$$V_{sol} = [(r_m^{g/s})^2 - (r_m^{g/v})^2]h\pi\frac{\varphi}{360} , \tag{3}$$

with the indices: gel = gel phase, sol = solvent phase, b = bottom, m = meniscus, 0 = initial and the phase boundaries: g/v = gel vapor, s/v = solvent/vapor, g/s = gel/solvent.

With the density/concentration relation the initial density of the gel ρ^0 can be calculated.

$$\rho^0 = (aw_2^0 + b)^{-1} . \tag{4}$$

If ρ^0 is known, we can determine the masses m of the various phases:

$$m_{gel}^0 = V_{gel}^0 \rho^0 , \tag{5}$$

$$m_{sol} = V_{sol}\rho_{01} , \text{ and} \tag{6}$$

$$m_{gel} = m_{gel}^0 - m_{sol} . \tag{7}$$

The average density of the gel in the equilibrium case ρ is given by the ratio m_{gel}/V_{gel}. With the values which can be calculated by use of Eqs. (1)—(7) the mass of the polymer in the gel m_2, the average concentration of the polymer in the gel phase $\overline{w}_2$, and the density of the gel at the cell bottom ρ_b can be obtained:

$$m_2 = m_{gel}^0 w_2^0 \tag{8}$$

$$\overline{w}_2 = \frac{m_2}{m_{gel}} . \tag{9}$$

Although we know that a linear density gradient-distance curve leads to a parabolic density-distance relation, it has been assumed that the density depends linearly upon the distance. This was assumed by Svedberg and experimentally shown by Johnson [1—3]

$$\rho_b = 2\rho - \rho_{g/s} . \tag{10}$$

The mean density of the gel ρ is the arithmetical mean of the gel density at the bottom of the cell ρ_b and the density of the meniscus gel/solvent $\rho_{g/s}$.

At the meniscus gel/solvent the mass fraction of the polymer in the gel is that of the maximum swollen gel $w_{2,s}^0$, which can be obtained separately from gravimetric swelling experiments. With the help of $w_{2,s}^0$ the polymer concentration at the cell bottom $w_{2,b}$ can be calculated:

$$w_{2,b} = 2\overline{w}_2 - w_{2,s}^0 . \tag{11}$$

Now it is possible to determine the slope a_1 and the intercept b_1 of the above mentioned linear function $\rho(r) = a_1 r + b_1$ which describes the dependence of the density of the gel on r. This is obtained by inserting the corresponding quantities at the meniscus gel/solvent and the bottom of the cell

$$a_1 = \frac{\rho_b - \rho_{g/s}}{r_b - r_m^{g/s}} \text{ and} \tag{12}$$

$$b_1 = \rho_b - a_1 r_b . \tag{13}$$

In analogy to $\rho(r)$ the function $w_2(r) = a_2 r + b_2$ is assumed to be linear in a small concentration range which leads to

$$a_2 = \frac{w_{2,b} - w_{2,s}^0}{r_b - r_m^{g/s}} \quad \text{and} \tag{14}$$

$$b_2 = w_{2,b} - a_2 r_b . \tag{15}$$

When the equilibrium is reached the swelling pressure for a deformed isotropic two component gel for every polymer concentration at every point in the external field can be calculated from the data of the Schlieren patterns with Eq. (19e) of part I. In addition to the equation which was used before [5] it is now possible to calculate the swelling pressure for higher gel concentrations, too [7, 8]:

$$\Pi_s = -\omega^2 \int_{r_m^{g/s}}^{r} \frac{r}{\tilde{V}_{01}} [1 - \tilde{V}_{01} \rho(r)] dr . \tag{16}$$

The quantities are: Π_s = swelling pressure, ω = angular frequency, $\tilde{V}_{01}$ = specific volume of the pure solvent, $\rho(r)$ = local density of the gel, and r = radial distance from the axis of rotation (index m is the meniscus, g/s stands for gel/solvent).

With the integrated form of Eq. (16) it is possible to calculate the swelling pressure Π_s for every distance r. The corresponding concentration in the mass fraction scale can be derived by means of Eqs. (14) and (15).

$$\Pi_s = \omega^2 \left(\frac{1}{3} a_1 r^3 + \frac{1}{2} b_1 r^2 - \frac{1}{2} \rho_{01} r^2 \right.$$

$$- \frac{1}{3} a_1 (r_m^{g/s})^3 + \frac{1}{2} b_1 (r_m^{g/s})^2$$

$$\left. - \frac{1}{2} \rho_{01} (r_m^{g/s})^2 \right) \tag{17}$$

With this equation the swelling pressure Π_s as a function of the polymer concentration in the range from the minimum at the meniscus gel/solvent to the maximum concentration of the polymer at the bottom of the cell has been calculated. If at the meniscus gel/solvent the gel is in contact with the pure solvent, the swelling pressure must be equal to zero.

If we calculate the differential of Eq. (16) we are able to control whether the values for the swelling pressure Π_s are determined properly or not. At the meniscus gel/solvent Π_s is equal to zero and,

therefore, the value of the derivative $\partial \Pi_s / \partial r$ at $r = r_m^{g/s}$ is used for the control

$$\left(\frac{\partial \Pi_s}{\partial r} \right)_{r=r_m^{g/s}} \sim \omega^2 r_m^{g/s} . \tag{18}$$

A further proof that Π_s is calculated correctly can be achieved if $\left(\dfrac{\partial \Pi_s}{\partial w_2} \right)_{w_2 = w_{2,s}^0} \left(\dfrac{\partial w_2}{\partial r} \right)_{r=r_m^{g/s}}$ is constant for different rotational speeds if all other conditions are kept constant.

3. Materials

Gelatin is a polypeptide derived from collagen by means of denaturation. According to the manufacturing process there are two types of gelatin: type A, which is derived from pigskins, and the alkali-trated gelatin type B from the raw material of bones. A photographic gelatin type B with $M_n = 68000$ g/mol, a water content of 14.2% by wt., and a broad distribution of molar masses is used for these investigations. The samples of gelatin/water gels are in a range of initial polymer concentrations from 2.94 to 10.62% by wt.

The linear polysaccharide κ-carrageenan has been extracted from seaweed by alkali treatment. Carrageenan as a natural product is a mixture of different polysaccharides; therefore, the κ-type is not derived purely [9]. The potassium salt of κ-carrageenan with a water content of 12.2% by wt. and a potassium content of 10.0 ± 0.5% is investigated. κ-carrageenan gives a rigid gel with water. The initial polymer concentrations of the gels varied in a range from 0.99 to 3.83% by wt.

4. Experimental

A non-dialyzed industrial gelatin was dissolved in distilled water. Afterwards, 0.15 ml of a methanol solution containing 5% by wt. chlorophenol was added per g gelatin granules. The granules were allowed to swell at least for 2 days in a refrigerator at about 7°C. The swollen sample was stored in a refrigerator. To prepare a sample for the experiment the gelatin was heated to 45°C for 60 min. During this time the bottle was shaken a few times.

The fine powdered κ-carrageenan was treated in the same way, but heated to 80°C for 1—2.5 days.

The maximum degree of swelling of the gels needed to calculate the swelling pressure-concentration curves is determined in analogy to that described in part II [8].

Before mounting the cells the walls of the centerpieces of the ultracentrifugal cells have been prepared with a thin film of silicon oil of high viscosity or teflon to avoid adhesion of the gel at the cell walls. The windows have been prepared with a thin film of silicon oil.

About 0.4 ml of the hot polymer solution was filled into the centerpiece which was in such a position that the gravitational field acted in the same direction as the later centrifugal field. The samples were allowed to gel for 4 days at the desired gelling temperature. In case of enclosed air bubbles in the gel the mounted and filled cell was heated up in the mentioned way until the air bubbles disappeared. The polymer solution was brought to gelation afterwards.

The rotor was conditioned in the rotor chamber for the gelling time at the chosen gelling temperature of the gels, then the cells were placed into the rotor quickly. After conditioning of the rotor for 1 more h the run was started.

The rotor was accelerated to the selected rotational speed. If the Schlieren patterns did not change anymore in the period of 24 h the equilibrium sedimentation of the gel was considered to be reached. Subsequently, the rotational speed was raised by 4000 rotations per minute (RPM) for gelatin and 6000 RPM for κ-carrageenan until a distinct movement of the meniscus gel/solvent in the direction of the cell bottom became obvious. Afterwards, the velocity was again lowered to the former selected speed; after the new equilibrium was reached the run was finished.

5. Results and discussion

It could be shown that Eq. (18) is fulfilled and that

$$\left(\frac{\partial \Pi_s}{\partial w_2}\right)_{w_2=w_{2,s}^0} \left(\frac{\partial w_2}{\partial r}\right)_{r=r_m^{g/s}} \text{ is a constant for all}$$

investigated gels. Therefore, the swelling pressure has been calculated properly due to the used approximation.

5.1. The system gelatin/water

To calculate swelling pressure curves from the data of an ultracentrifugal experiment it has to be confirmed that the sedimentation of the gel leads to a real equilibrium deformation. This can be demonstrated by means of the relative deformation r^* of the gel, which is defined by

$$r^* = \frac{r_m^{g/s} - r_m^{g/v}}{r_b - r_m^{g/v}}, \tag{19}$$

with the indices: g/s = gel/solvent, g/v = gel/vapor, m = meniscus, and b = cell bottom.

This relative deformation can be plotted vs time of sedimentation at a given rotational speed. In this way two branches are obtained, including the equilibrium value, either from zero RPM (lower branch) or from higher speeds (upper branch) [4, 5]. This procedure only works if there are no superpositions of different equilibria with the swelling pressure equilibrium or irreversible phenomena like crystallization, changes of the crosslinking density, etc.

In contrast to the experiments with a gelatin, where no soluble parts could be detected, in this contribution results are presented which clearly show that a superposition of a sedimentation equilibrium of the gel and that of the soluble polymer in water is given.

In case of gelatin/water where during the gelation process not all of the polymer was incorporated into the network, we observed that the swelling pressure-concentration curves did not superimpose for different rotational speeds, as was the case for a polymer where no soluble part could be detected. This is demonstrated in Fig. 2, where the swelling pressure-concentration curves of the same initial concentration, but different rotational speeds are shown.

The result is controversal to the predictions of Eq. (16) which will give only a single swelling pressure-concentration curve if the rotational speed is changed [8]. It is supposed that during the time allowed for sedimentation true equilibria have not yet been reached caused by the presence of soluble polymer material. The explanation may be given that there is an influence of the soluble parts which increases with higher rotational speeds. The polymer remaining in the sol after gelation will establish a sedimentation-diffusion equilibrium which is superimposed to the swelling pressure equilibrium of the gel. Thus, the activity of the solvent is changed which may lead to the higher deswelling of the gel at the same swelling pressure. Because of the much higher molar mass of the gel compared with that of the soluble parts the swelling pressure equilibrium of the gel is achieved much quicker than the sedimentation-diffusion equilibrium of the soluble parts. Therefore, if the rotational speed is lowered from 22000 RPM to 18000, the gel does not swell to such an extent as before when the same rotational speed of 18000 RPM was reached from 0 RPM,

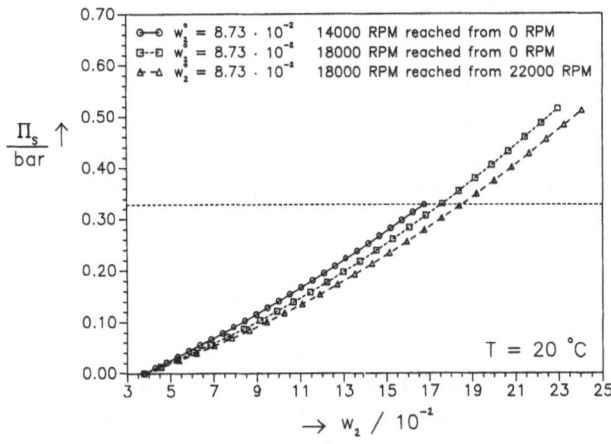

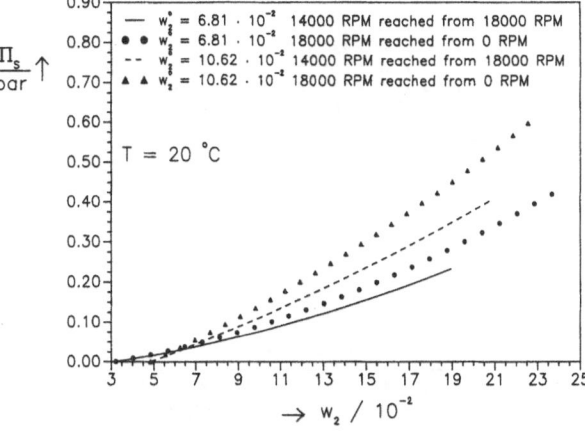

Fig. 2. Swelling pressure Π_s of the system gelatin/water vs concentration of gelatin for different rotational speeds; w_2 is the mass fraction of the polymer, w_2^0 its initial value

Fig. 3. Swelling pressure Π_s vs concentration for the system gelatin/water for different rotational speeds and initial concentrations; w_2 is the mass fraction of the polymer, w_2^0 its initial value

because the gradient of the soluble parts corresponds to the higher rotational velocity. It seems that the true equilibrium corresponding to the superposition mentioned is not yet reached.

To achieve a rough estimation of the time needed for the sedimentation-diffusion equilibrium of the soluble parts to be reached, we used the definition of Van Holde and Baldwin of the time t_δ which is needed to reach equilibrium [10]. In our estimation we noticed the restrictions for satisfactory sedimentation measurements made by Fujita [11]. For the Beckman 4° monosector cell with a column heigt of 10 mm, we yield times of 35—40 days until the equilibrium of the soluble parts will be reached. It is clear that the calculation of t_δ is not very correct, because we do not know the sedimentation and diffusion coefficients of the soluble parts in the gel phase which have been roughly estimated.

In Fig. 2 it can be seen that the swelling pressure-concentration curves differ a little for different rotational speeds. The same swelling pressure as indicated by a horizontal dashed line leads to higher polymer concentrations if the rotational velocity is increased. This is equivalent to the mentioned deswelling of the gel if the gradient of the soluble parts becomes steeper in the gel phase after the rotational speed is increased. Therefore, we cannot treat this system as binary as we did in the calculations before [4—6, 8]. For a dialyzed photographic gelatin type A without soluble parts the existence of a sedimentation equilibrium was proven by Holtus [4, 5].

The extraordinary phenomena provoked by soluble parts are observed for different initial concentrations. The swelling pressure-concentration curves for 6.81 and 10.62% by wt. of gelatin are presented in Fig. 3.

It can be seen that the curves for different rotational speeds, but the same initial concentration and the same degree of swelling at the boundary

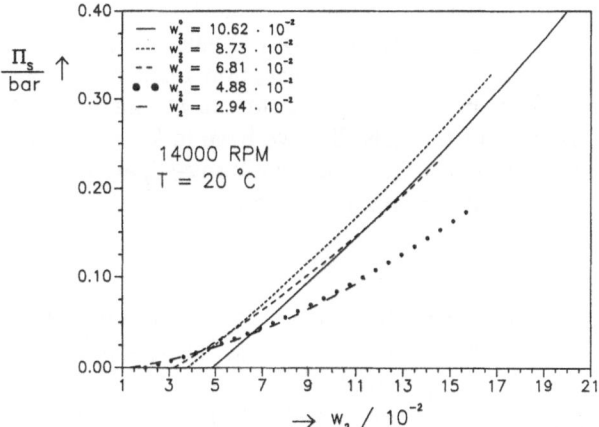

Fig. 4. Swelling pressure Π_s of the system gelatin/water vs concentration of gelatin for five different initial concentrations at 14000 RPM; w_2 is the weight fraction of the polymer, w_2^0 its initial value

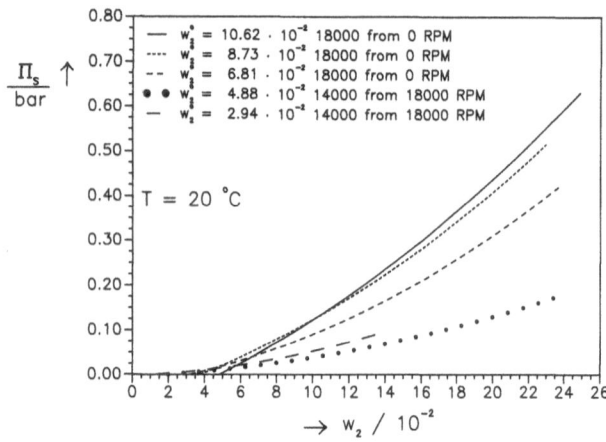

Fig. 5. Swelling pressure Π_s of the system gelatin/water vs concentration of gelatin for five different initial concentrations at 18000 RPM, w_2 is the weight fraction of the polymer, w_2^0 its initial value

gel/sol do not coincide. The curves with the "preshistory" of a higher preceding rotational speed are situated below those where the speed is attained by a start from 0 RPM. This difference has to be explained by a closer look to the soluble part of the gels in the future.

A further problem is stated if we are considering swelling pressure-concentration curves, where the rotational speed for different initial concentrations is 14000 RPM. The result is shown in Fig. 4 for five different w_2^0-values. As in the preceding paper II, these curves are intersecting if the polymer concentration is below a certain overlap limit.

This seems to be valid for the values higher than $w_2^0 = 6.8 \cdot 10^{-2}$. The explanation for a binary gel can only be given by arguments of the network structure of the gels. But we have to keep in mind that in the non-dialyzed gelatin investigated in this paper soluble parts are present.

If the rotational speed is now increased to 18000 RPM all swelling pressure-concentration curves intersect, which can be derived from Fig. 5.

Once more, this can be related to the presence of soluble parts. The concentration gradient of the soluble polymer causes the gels to deswell additionally to the influence of the outer field, so that we have to be very careful in the interpretation of intersecting swelling pressure-concentration curves. We state that there are two different arguments for this effect: a) inhomogeneities in the network structure and b) the presence of additional gradients of soluble polymer in the network.

A further effect of the superposition of the gradient of the soluble parts with the swelling pressure equilibrium is the influence of the initial gel volume in the measuring cell on the swelling pressure curves. Because the initial gel volume influences the gradient of the soluble parts the swelling pressure curves do not coincide if gels with only varied initial volume are measured under the same conditions. From the theory it is expected that the swelling pressure curves for gels with different initial volumes coincide. The swelling pressure-concentration curve of the gel with the higher initial volume extends to higher concentrations than that of the gel with the lower initial one.

The chemical composition of the soluble parts and their molar mass distribution are not yet known. The quantity of soluble parts seems to be independent of the time of gelation, as shown in Fig. 6.

The concentration gradient of the soluble parts is nearly the same for all gelling times. Thus, it seems that the soluble parts have no crosslinking ability. Therefore, they cannot be built into the network. It may be supposed that the soluble parts are of low molar mass. This seems to be possible because the collagen is highly fragmented by an alkaline treatment during the production process of gelatin type B.

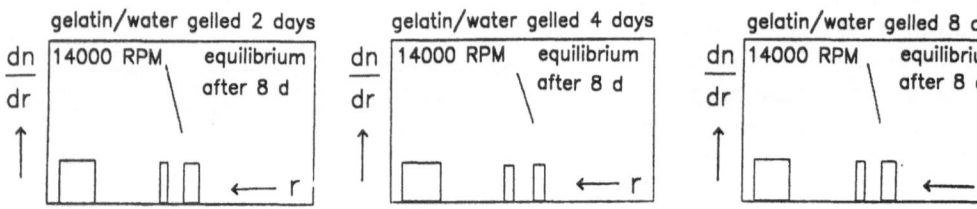

Fig. 6. Schlieren patterns of the sol regions of samples of gelatin/water measured under the same conditions. The gelling time was varied between 2 and 8 days. The initial concentration is 6.81% by wt. of gelatin in all cases. r is the radial distance from the axis of rotation

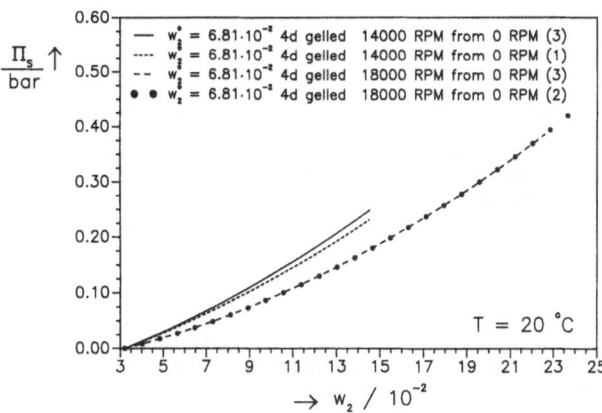

$\frac{\Pi_s}{bar} \uparrow$

— $w_2^* = 6.81 \cdot 10^{-2}$ 4d gelled 14000 RPM from 0 RPM (3)
---- $w_2^* = 6.81 \cdot 10^{-2}$ 4d gelled 14000 RPM from 0 RPM (1)
-- $w_2^* = 6.81 \cdot 10^{-2}$ 4d gelled 18000 RPM from 0 RPM (3)
●● $w_2^* = 6.81 \cdot 10^{-2}$ 4d gelled 18000 RPM from 0 RPM (2)

T = 20 °C

$\rightarrow w_2 / 10^{-2}$

Fig. 7. Swelling pressure Π_s of the system gelatin/water vs concentration of gelatin for different rotational speeds in different experiments (1—3); w_2 is the weight fraction of the polymer, w_2^0 its initial value)

Although the method of measuring swelling pressures with an analytical ultracentrifuge is very sensitive the swelling presure curves of gels which are measured under the same conditions including the prehistory nearly coincide as shown in Fig. 7 for the system gelatin/water in 3 different experiments (1—3).

It can be stated that the experimental error is small and the reproducibility of the measurements is very good. Therefore, the effects described in this article are out of experimental errors.

5.2. The system κ-carrageenan/water

In the case of κ-carrageenan/water, it is possible to demonstrate equilibria of the continuous system by means of r^*. It can be seen from Fig. 8 that the two branches for reaching the equilibrium from different sides approach with time. Because it is ineffective to wait in each case until both branches show the same value, the average between both values of the meniscus gel/solvent is used for further calculations if the meniscus does not obviously move anymore.

Considering the swelling pressure-concentration curves of gels with different initial concentrations for the system κ-carrageenan/water, it can be seen that those of the gels with initial concentrations of 2.90 and 3.83% by wt. carrageenan do not intersect, which can be seen in Fig. 9. This means that the gel may be considered as a homogeneous network in this concentration range.

Also, in this case the equilibrium degree of swelling has been determined by separate gravimetric measurements.

Although soluble parts are present (which is demonstrated in Fig. 10 by gradients of the refractive index in the sol located to lower r-values than those of the gel), they apparently do not influence the swelling pressure equilibria here. Thus, it is possible to prove the existence of equilibria by means of r^* for the system κ-carrageenan/water.

The discontinuous steps in the refractive index gradient dn/dr indicate irreversible aggregation of

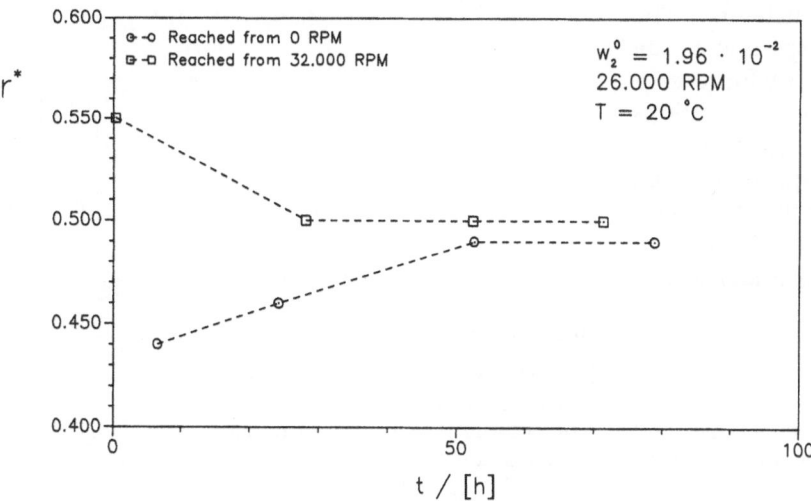

r^*

⊖-⊙ Reached from 0 RPM
⊟-⊡ Reached from 32.000 RPM

$w_2^0 = 1.96 \cdot 10^{-2}$
26.000 RPM
T = 20 °C

t / [h]

Fig. 8. Movement of the relative meniscus gel/solvent r^* vs time t for the system κ-carrageenan/water (w_2 is the weight fraction of the polymer, w_2^0 is its initial value) after having reached the speed of 26000 rotations per minute (RPM) from different starting condions

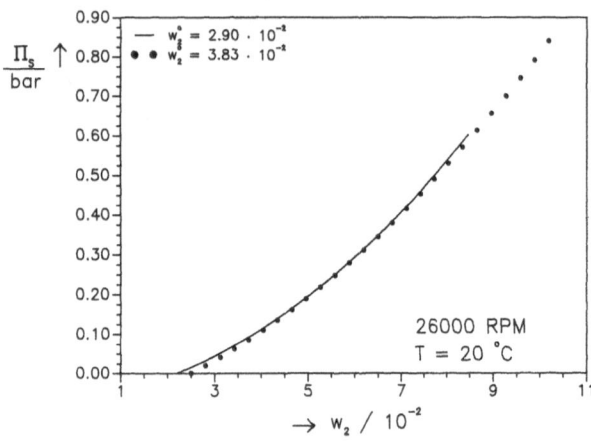

Fig. 9. Swelling pressure Π_s of the system κ-carrageenan/water vs concentration of κ-carrageenan for different initial concentrations; w_2 is the mass fraction of the polymer, w_2^0 its initial value

References

1. Svedberg T, Pedersen KO (1940) Die Ultrazentrifuge, Steinkopff Dresden
2. Johnson P (1971) J Photograph Sci 19:49
3. Johnson P (1970) Velocity and equilibrium aspects of the sedimentation of agar gels in Photographic Gelatin I, Editor Cox RJ, Academic Press, London:13
4. Holtus G, Borchard W (1989) Colloid Polym Sci 267:1133
5. Holtus G (1990) Dissertation, Duisburg
6. Cölfen H (1991) Diplomarbeit, Duisburg
7. Borchard W this issue
8. Cölfen H, Holtus G, Borchard W this issue
9. Therkelsen GH, private communication
10. van Holde KE, Baldwin RL (1958) J Phys Chem 62:734
11. Fujita H (1962) Mathematical Theory of Sedimentation Analysis, Academic Press, New York and London:300

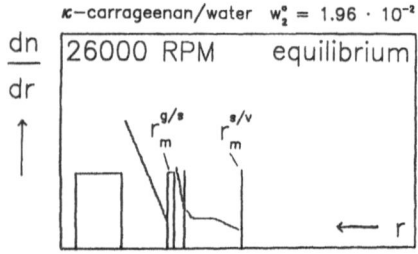 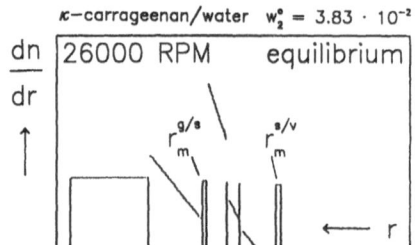

Fig. 10. Schlieren patterns of κ-carrageenan gels with associated soluble parts and different initial concentrations in the equilibrium case (see text)

the polymer and the formation of microparticles during the sedimentation. Furthermore, the aggregates do not move into the network. The phenomena observed in this experiment are perhaps described by the Gilbert theory for reversible polymerizing systems if additional conditions are introduced [12, 13]. Possibly, the aggregates are also formed by other types of carrageenan which are soluble impurities of the κ-type like λ-, ξ, and θ-type which have no gelling abilities [9]. But, in general, this phenomenon is not yet understood.

12. Gilbert GA (1955) Disc Farad Soc 20:68
13. Gilbert GA (1959) Proc Roy Soc (London) A250:377

Acknowledgements

We thank the Deutsche Forschungsgemeinschaft and the Max Buchner-Forschungsstiftung for financial support of the project.

Authors' address:

Prof. Dr. W. Borchard
Universität-GH-Duisburg
Angewandte Physikalische Chemie
Lotharstr. 1
4100 Duisburg 1, FRG

Progress in Colloid & Polymer Science   Progr Colloid Polym Sci 86:111—118 (1991)

Future requirements for modern analytical ultracentrifuges

W. Mächtle

Kunststofflaboratorium, Abt. Polymerphysik, Festkörperphysik
BASF Aktiengesellschaft, Ludwigshafen/Rhein, FRG

Abstract: The last improvements on the most used analytical ultracentrifuge (AUC), the Model E from Beckman Instruments, were made 30 years ago. Only a few individual laboratories have made some further improvements. There has been, indeed, an anouncement of a new AUC from Beckman for 1991, with some interesting refinements, but we think it is possible to do it better. Thus, we propose details of future AUC. Some of the proposed features are already realized in individual labs, some have to be developed in the future. The proposed main features are the following: 1) Multi-hole-rotors; 2) Multiple detection systems; 3) Automatic on-line data analysis, and 4) A new running technique with continuously increasing rotor speed ω.

Key words: Analytical ultracentrifuge; molar mass distribution; particle size distribution; Schlieren optics; video camera; image-processing; multiple detection; macromolecules; dispersions

1. Introduction

Everybody in the world of analytical ultracentrifugation is happy about the announcement [1] of a new analytical ultracentrifuge for 1991, the Optima XL-A from Beckman. The announcement is timely after 30 years of stagnation in the development of the Spinco-Beckman Model E, the most used AUC in the world.

We think it is helpful to discuss at the beginning of a new era of ultracentrifugation how a modern AUC should look like. The new Optima XL-A has some interesting refinements, but not enough to be better than the old Model E in all aspects. Thus, our following proposals are directed mainly to all development teams of AUC. Some of our proposed features are already realized in individual labs, some have to be developed in the future. In this paper, we would like to discuss four main requirements for a modern AUC:

1) Multi-hole-rotors;
2) Multiple detecting systems;
3) Automatic on-line data analysis, and

4) A new running technique with a *variable* rotor speed ω during a run.

The great potential inherent in the AUC technique to analyze and fractionate macromolecules and, especially, microparticles (colloids) is only been partly realized. The realization of our proposed four requirements in one instrument will allow to use the full potential of the AUC.

2. Requirement 1: Multi-hole-rotors

To simultaneously measure several samples instead of only one is very efficient. Further, it improves the accuracy of measurement and allows better comparisons. To measure as many samples as possible in one run requires multi-hole-rotors.

Some labs in Germany, like ours [2], have used eight-hole-rotors for many years to measure eight samples simultaneously. Figure 1 shows our modification of a Beckman Model E with a trigger device, a multiplexer, and a triggerable light source: a laser or a flash lamp [3]. Figure 2 shows a meas-

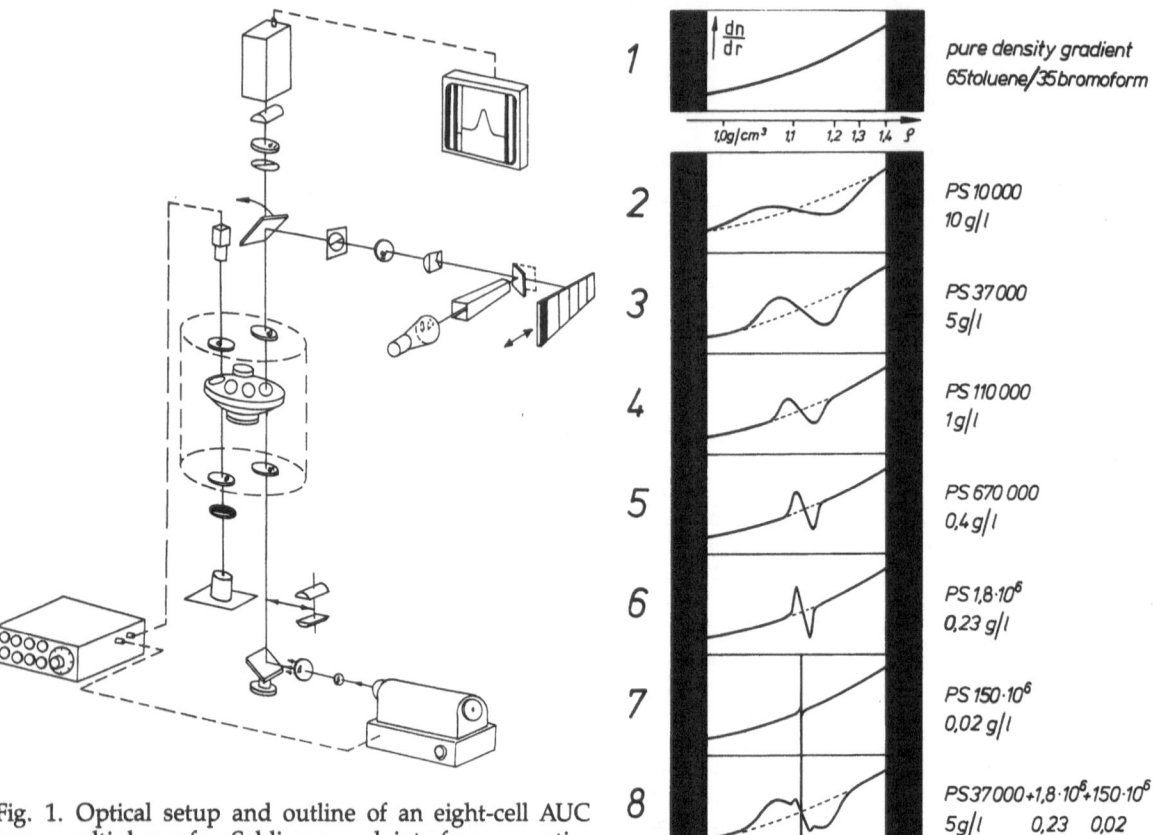

Fig. 1. Optical setup and outline of an eight-cell AUC laser multiplexer for Schlieren and interference optics with photo plate and digital TV camera

Fig. 2. Eight-cell density gradient run of seven different polystyrene samples, measured simultaneously in one night with an AUC Schlieren optics multiplexer. A 65 toluene/35 bromoform (wt.%)-density, gradient is used. The abscissa below the upper Schlieren photo shows the density ρ inside the cell as a function of the rotor radius. The ordinate shows the refractive index gradient dn/dr

uring example from this modified Model E, eight density gradient Schlieren photos of different polystyrene samples in eight different cells measured simultaneously in one run. The time needed to reach equilibrium was as long as 24 h. That means that this valuable analytical tool — the AUC-density-gradient technique — can be used efficiently only with an eight-cell-rotor.

We think Beckman has to offer us such an eight-hole rotor, too, instead of the announced four-hole rotor. Eight holes is probably the maximum number for a maximum rotor speed of 60 000 rpm because of material reasons. But the maximum speed of 60 000 rpm is very seldom needed. Nearly all of our sedimentation runs are performed with 40 000 rpm because we use 30-mm-cells in most cases. For the particle size distribution measurements [4, 5] which are very important for industry, a maximum rotor speed of 20 000 rpm is sufficient and, for nearly all equilibrium runs to measure molar masses, 10 000 rpm are enough.

Thus, beside the eight-hole high-speed rotor, we need a low-speed analytical rotor, but one with 20, or better, up to 50 holes. We think Beckman is able to realize this. The 20 to 50 cells could be arranged in two or more circles around the rotor axis like in preparative rotors. New cells are needed to realize this. Especially for particle size distribution measurements it is conceivable to cut out sector shaped holes directly into the rotor core.

3. Requirement 2: Multiple detection systems

The new Optima XL-A will offer us only one detector, a UV-scanner. That is not enough. An UV-scanner is useful for biopolymers and some industrially important polyelectrolytes. But it is not useful for synthetic polymers which require UV-absorbing organic solvents.

Unfortunately, most AUC-development teams consider only biopolymers and forget about the much greater area of synthetic polymers, the huge world of plastic material, and, especially, the very important polymer latices and micro particulate systems of dyestuffs, laquers, and paints.

Each polymer lab working with GPC, liquid, and size exclusion chromatography is a potential user for an AUC. Therefore, we request of Beckman: beside the UV-scanner for the new Optima, we need additionally a refractive detection system, that means Schlieren and Rayleigh interference optics.

But besides UV-scanner and refractive detectors, we need more detectors in a modern AUC. Figure 3

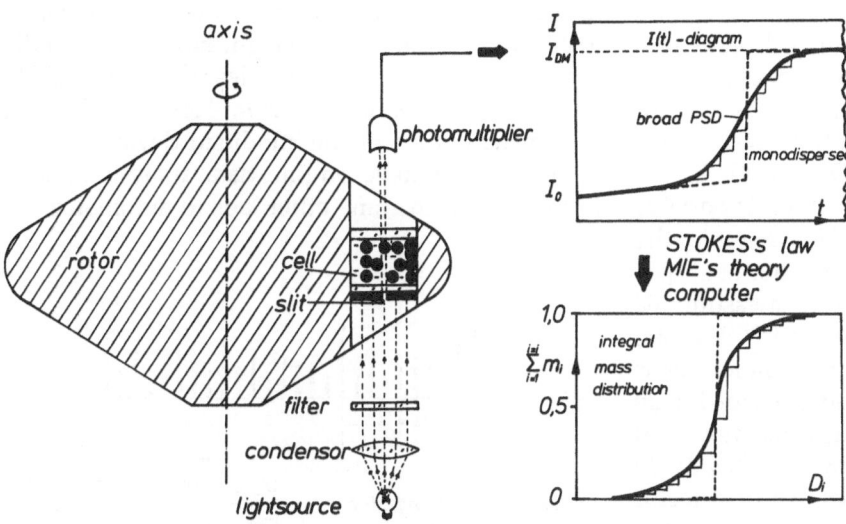

Fig. 3. Experimental AUC setup to measure the particle size distribution of a dispersion with a turbidity detector

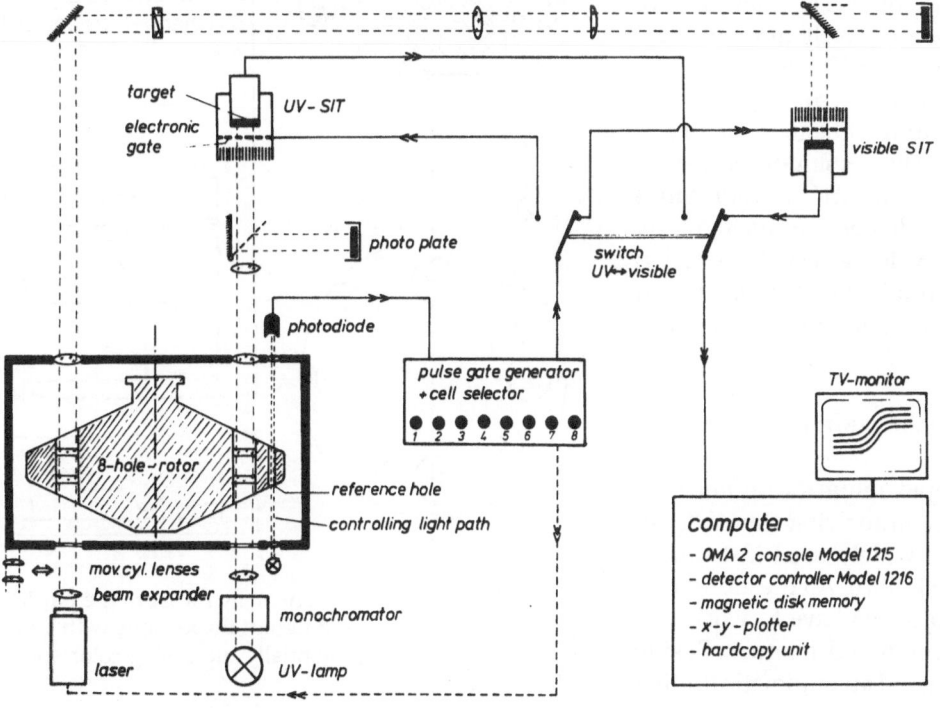

Fig. 4. Optical setup and outline of an AUC with an eight-cell-multiplexer, two optical paths, photo plate, optical multichannel analyzer, and four different detectors (UV-scanner, Schlieren optics-, interference optics- and turbidity-detector)

shows that to analyze microparticulate systems which reveal turbidity, like polymer latices and pigments, we need an additional detector, a turbidity or light-scattering detector [4] to calculate from primarily measured light-intensity I — running time t-curves by means of Stokes' law and Mie's light-scattering theory particle size distribution curves. For the analysis of polymer latices this simple turbidity detector is a valuable tool.

Figure 4 shows that, already in 1979, we had realized an AUC with four different detection systems [6], working in the two optical paths of a Model E: a UV-scanner-, a Schlieren optics-, and interference optics-, and a turbidity detector. Today, we think Beckman should be able to realize this four-detector system in the new Optima XL-A, too.

A four-detector system is useful. But it is possible to add more interesting detectors, for example, the excellent fluorescence detector of Riesner et al. [7]. This fluorescence detector is able to measure 100 times smaller concentrations than the UV-Scanner, if a fluorescence labeling of the macromolecules is possible. Furthermore, an infrared detector inside an AUC would be marvelous, but it has not been realized until now; it would allow to differentiate between different chemical species. An IR-detector would multiply the analytical possibilities of the AUC.

We emphasize in connection with this statement: The ultracentrifuge is a *fractionation* instrument and it allows to fractionate macromolecules and microparticles according to *two*, completely different fractionation principles: first, according to size in the sedimentation run, and second, according to density, that means according to the chemical nature in the density gradient run. The resolution power according to molar mass of an AUC is four times higher than that of size exclusion chromatography (because of the elution volume V/sedimentation coefficient s-molar mass M-relations $V \propto M^{\sim 0.1}$ and $s \propto M^{\sim 0.4}$). Most people are not aware of this fact.

4. Requirement 3: Automatic on-line data analysis

The introduction of a rapid automatic on-line data analysis system is the key to reviving the AUC, to make it again a powerful, widely used instrument as it was in the period 1950—1970. All the aforementioned optical detection systems have to work in a rapid on-line-computer controlled manner. Beside the multi-hole rotor technique an automatic on-line

data analysis is the second measure required for the AUC.

Already in 1975, we realized such an automatic on-line data analysis in our AUC-turbidity detector device (Fig. 3) to automatically measure particle size distributions of polymer dispersions. The signals of the photomultiplier in Fig. 3 were instantly digitized and transferred to a computer. Thus, immediately after the AUC-run was finished the computer was able to plot the final particle size distribution diagram in the lower part of Fig. 3. Later we introduced [4] an eight-hole rotor into this device to measure seven samples simultaneously by that digital turbidity detector. Figure 5 shows a schematic outline of this modified device.

Similar automatic on-line data analysis systems were realized in some other labs to automatize the

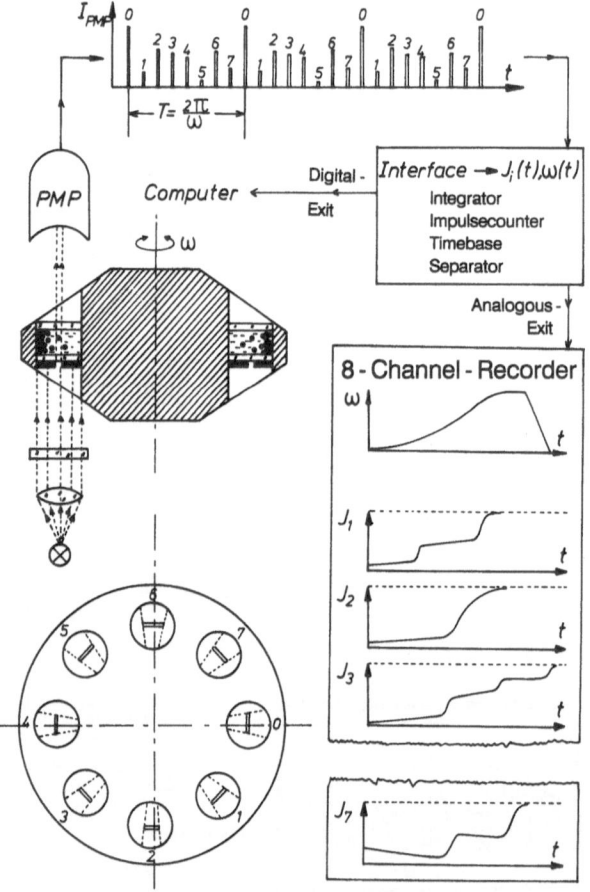

Fig. 5. AUC setup to simultaneously measure the particle size distribution of seven different dispersions with a turbidity detector and exponentially increasing rotor speed ω during the run

UV-scanner of the Beckman Model E. Also, the fluorescence detector of Riesner et al. [7] works in a digital manner. But an on-line data analysis for the Schlieren and the interference detector has not yet been realized, because it is much more difficult. The upper part of Fig. 1 shows our first steps towards the Schlieren optics [8]. We replaced the photo plate of the Model E by a digital TV-camera with 500 × 500 pixels. But for a better resolution we would like to have a TV camera with 5000 × 5000 pixels. We think that within the next decade the market will offer such high-resolution digital TV-cameras.

Figure 6 shows an automatic data analysis of our present digital Schlieren picture analysis system. The upper photo is the primary recorded TV-pic-

ture. The second photo shows the brightness corrected picture, and the third photo the reduced binary picture. The bottom photo shows the final reduced data set made up from 500 yellow marked pixels which are superimposed onto the primary recorded TV-picture. But, at present, our system is not a real on-line system. We record the TV-pictures during the run and process them in a computer after the run.

We are realizing an automatic Schlieren optics data analysis. Floßdorf [9] realized the same with interference optics. Laue [10] and Stafford [11] are developing interference optics systems. We hope Beckman will offer a Schlieren- and interference-optics on-line data system for the new Optima XL-A as soon as possible. For synthetic polymers such a detection system based on refractive index or refractive index gradient is very important.

The accuracy of sedimentation runs will be increased considerably, if we take 100 electronic pictures per run instead of only five old-fashioned photos. An AUC with a rapid digital on-line detection system based on refractive index is an interesting alternative to the size exclusion chromatography dominating at present in measuring molar mass distributions of polymers.

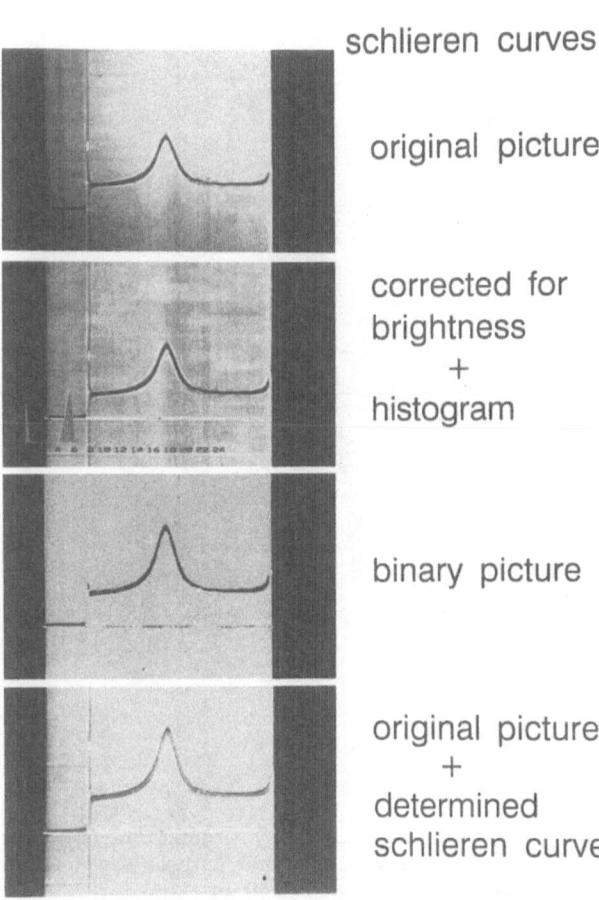

schlieren curves

original picture

corrected for
brightness
+
histogram

binary picture

original picture
+
determined
schlieren curve

Fig. 6. Automatic digitalization of a AUC-Schlieren picture with a digital 500 × 500 pixel-TV camera (see text). Sedimentation run of polyvinyl pyrrolidone in H_2O (c = 4 g/l, 30-mm monosector cell, ω = 40000 rpm, running time 224 min, philpot angle = 70°)

5. Requirement 4: New running technique (ω = variable)

In this section we propose a new AUC-running technique for sedimentation runs. Instead of a rotor speed ω being constant during a run we should use a variable rotor speed $\omega(t)$, i.e., in all sedimentation equations we replace the running time t, or more precisely, the value $\omega^2 \cdot t$, by the measured traveling time integral $\int \omega^2 dt$.

A precondition for realizing this is a completely new computer controllable maintenance-free rotor drive. That means a vacuum included induction motor without gear box and oil bearing. The new Optima XL-A has such a modern drive. The old drive of the Model E is not convenient for the proposed new running technique. The high costs of maintenance for the old Model E-drive (beside the very high prices for rotors and cells) are one reason for the decline of the AUC in the last 15 years.

A new, computer-controllable drive, i.e., a new running technique of exponentially increasing rotor speed during a run, instead of the old constant rotor speed offers some interesting advantages (Fig. 5).

We realized this technique 15 years ago for our particle-size distribution measuring technique with a turbidity detector by modifying the drive of a preparative ultracentrifuge [4]. The diagram inside of Fig. 5 shows rotor speed ω as a function of the running time t. In every run, in always the same manner, ω is increasing exponentially from 0 to a maximum rotor speed of 40000 rpm within 2 h. Thus, we never have to ask before a run, what is the most suitable constant rotor speed? Instead, in every run we measure the correct particle-size distribution with this exponential technique for very small 30-nm-particles and for large 3000-nm particles with a 10000 times higher sedimentation velocity.

As an example, Fig. 7 shows (for details see [4]) a comparison between the old running technique with a constant rotor speed of 4000 rpm and the new one, with an exponential increase from 0 to 40000 rpm. The upper part of Fig. 7 shows the relative light intensity $I_r = I/I_{LM}$ vs running time t fractionation curves of a polystyrene latex 40:50:10 wt.%-mixture with three different diameters. In both running techniques we see three I-t-steps, one of the fast 794 nm-particles, one of the 312 nm particles, and one of the slow 176 nm-particles.

In the center part of Fig. 7 the same $I(t)$-measuring curves of the upper part are plotted in a different form, i.e., the light intensity I is now plotted as function of the traveling time integral $\int \omega^2 dt$. This integral was calculated in a numerical manner by measuring the rotor speed ω every 5 s. The light intensity-traveling time integral-curves of both running techniques, $\omega =$ constant and $\omega =$ variable agree. We always see 14 curves in the diagrams of Fig. 5, because we measured the same sample in seven different cells simultaneously.

If the computer transfers these 14 I-$\int \omega^2 dt$-curves in particle size distribution curves, we get as a result the 14 curves in the lower part of Fig. 7, where the mass percentage m is plotted as a function of the particle diameter D. All 14 curves, for the old running technique, $\omega =$ constant and the new one, $\omega =$ variable, agree. Within $\pm 5\%$ we reproduce the known diameters of the three different components and the given mixing ratio. This is a proof that our new running technique with a variable quantity ω works and that our computer calculates correct traveling time integrals $\int \omega^2 dt$.

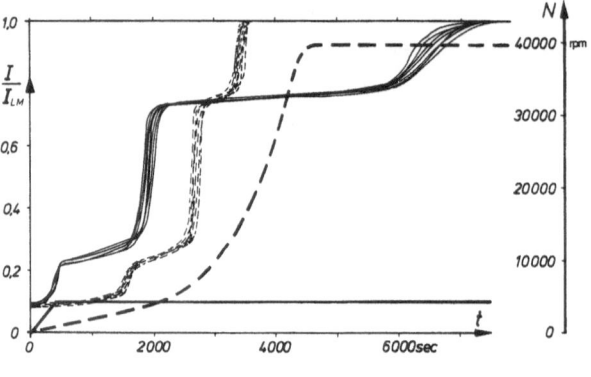

a

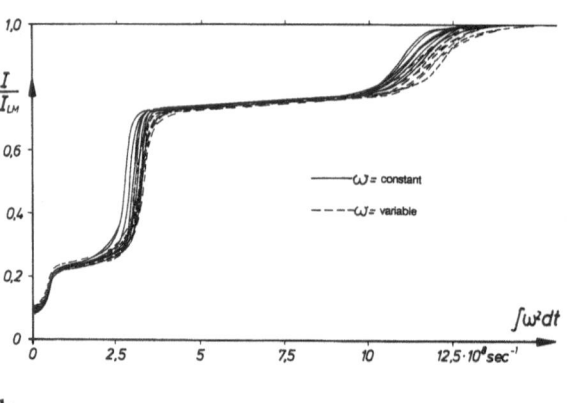

b

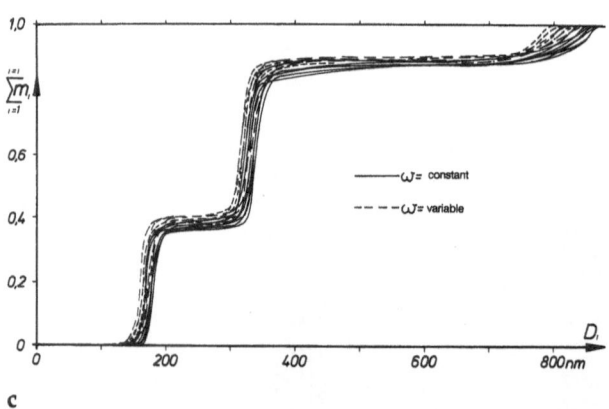

c

Fig. 7. Relative light intensity (I)-running time (t)-curves (top), I-$\int \omega^2 dt$-curves (center) and particle size distribution curves (bottom) of a 40/50/10 wt.-%-mixture of three monodisperse DOW polystyrene calibration latices having diameters of $D = 176/312$ and 794 nm. Seven-cell simultaneous run of the same 1 g/l-concentration: a) with constant rotor speed $\omega = 4000$ rpm (———) b) with variable rotor speed, i.e., roughly exponentially increasing from $\omega = 0$ to $\omega = 40000$ rpm (– – –) within 80 min

We think it is adventageous to apply this variable rotor speed technique, not only to high resolution particle size distribution measurements of very broad distributed polymer dispersions [12], but to all kinds of sedimentation runs of dissolved macromolecules, too, that means for Schlieren optics-, interference optics-, and UV-scanner runs.

In combination with synthetic boundary techniques we see interesting new analytical possibilities in the near future: first, in measuring absolute molar masses M by means of Svedberg's equation

$$M = \frac{R \cdot T}{1 - (\bar{v})\rho} \frac{s}{D^*},$$

with R = universal gas constant;
  T = absolute temperature;
  $(\bar{v})$ = partial specific volume of the solute;
  ρ = density of the solution;
  s = sedimentation coefficient of the solute;
  D^* = diffusion coefficient of the solute,

in one single run, and second, in measuring complete molar mass distributions of broadly distributed synthetic polymers. The idea behind this is that, at the beginning of such an "exponential" sedimentation run, at very low rotor speed, the time spreading of a Schlieren peak in a synthetic boundary experiment is controlled only by the distribution of diffusion coefficient D^* of the macromolecules. Thus, we can measure this diffusion coefficient distribution. But in the later part of this sedimentation experiment, at high rotor speed, the spreading and migration of the Schlieren peak is mainly controlled by the distribution of sedimentation coefficient s, i.e., by the molar mass distribution. Therefore, in the later part of this experiment we measure this sedimentation coefficient distribution. Both distributions (of D^* and s) together yield the molar mass distribution.

The new Optima XL-A drive, in combination with the variable rotor speed and the synthetic boundary technique promises to solve the very old problem of separation between diffusion and sedimentation within a sedimenting Schlieren peak.

6. Concluding remarks

We tried to show that there are four important future requirements for a modern AUC:

— Multi-hole-rotors;
— Multiple detection systems;
— Automatic on-line data analysis, and
— A new running technique (ω = variable).

There are some other requirements for a modern AUC which we will not discuss in this paper. But they shall be mentioned briefly:

— a good temperature stabilization is important;
— the AUC should be small in size, noise and maintenance free;
— the AUC should be composed in a simple modular manner with two to four optical paths for researchers who like to introduce their own modifications;
— new cells, especially for the synthetic boundary technique, and
— the new AUC should be economical.

If all of the aforementioned requirements for a modern AUC or a great part of them are realized in one instrument, we are sure that there will be a renaissance of this versatile fractionation instrument with its many different analytical techniques. The full analytical power of the AUC was used in the past only for biopolymer but never for synthetic macromolecules and, especially, not for colloidal systems, for which the AUC is predestined. A modern AUC will also allow to characterize synthetic macromolecules and colloids on a large scale.

References

1. Schachmann HK (1989) Nature 341:259—260
2. Mächtle W, Klodwig U (1979) Makromol Chem 180:2507—2511
3. Mächtle W, Klodwig U (1976) Makromol Chem 177:1607—1612
4. Mächtle W (1984) Markomol Chem 185:1025
5. Müller HG (1989) Colloid Polymer Sci 267:1113—1116
6. Mächtle W (1979) Preprints short communications IUPAC MAKRO, Mainz, 17.—21. Sept. 1979, Vol. II, pp 731—734
7. Schmidt B, Rappold W, Rosenbaum V, Fischer R, Riesner D (1990) Colloid Polymer Sci 268:45—54
8. Klodwig U, Mächtle W (1989) Colloid Polymer Sci 267:1117—1126
9. Flossdorf F, Gesellschaft für Biotechn Forschung, Braunschweig, FRG, private communication

10. Laue TM, University of New Hampshire, USA, private communication
11. Stafford WF, Muscle Research, Boston, Biomed Res Inst, USA, private communication
12. Mächtle W (1988) Angew Makromol Chem 162:35—52

Author's address:

W. Mächtle
ZKM/P — G 201
Kunststofflaboratorium
BASF Aktiengesellschaft
6700 Ludwigshafen/Rhein, FRG

Author Index

Subject Index